Synthesis Lectures on Learning, Networks, and Algorithms

Series Editor

Lei Ying, ECE, University of Michigan–Ann Arbor, Ann Arbor, USA

The series publishes short books on the design, analysis, and management of complex networked systems using tools from control, communications, learning, optimization, and stochastic analysis. Each Lecture is a self-contained presentation of one topic by a leading expert. The topics include learning, networks, and algorithms, and cover a broad spectrum of applications to networked systems including communication networks, data-center networks, social, and transportation networks.

Zhiwei (Tony) Qin · Xiaocheng Tang ·
Qingyang Li · Hongtu Zhu · Jieping Ye

Reinforcement Learning in the Ridesharing Marketplace

Zhiwei (Tony) Qin
Lyft
San Jose, CA, USA

Xiaocheng Tang
Meta
San Mateo, CA, USA

Qingyang Li
DiDi
San Jose, CA, USA

Hongtu Zhu
University of North Carolina
Chapel Hill, NC, USA

Jieping Ye
Alibaba Group
Hangzhou, Zhejiang, China

ISSN 2690-4306   ISSN 2690-4314 (electronic)
Synthesis Lectures on Learning, Networks, and Algorithms
ISBN 978-3-031-59639-1   ISBN 978-3-031-59640-7 (eBook)
https://doi.org/10.1007/978-3-031-59640-7

This Springer imprint is published by the registered company Springer Nature Switzerland AG
The registered company address is: Gewerbestrasse 11, 6330 Cham, Switzerland

If disposing of this product, please recycle the paper.

Contents

Introduction

1

The emergence of ridesharing,[1] led by companies such as DiDi, Uber, and Lyft, has revolutionized the form of personal mobility. Per (FortuneBusinessInsights 2021), the global rideshare market size was US\$76.48 billion in 2020 and is projected to grow to a total market value of US\$242.73 billion by 2028. However, how to improve operational efficiency is a major challenge for rideshare platforms, e.g., long passenger waiting time (Smith 2019) and as high as 41% vacant time for ridesharing vehicles in a large city (Brown 2020). The success of ridesharing, from the perspectives of the platforms, drivers, and passengers, requires sophisticated optimization of all the integrated components that collectively deliver the services.

Reinforcement learning (RL) is a machine learning paradigm that trains an agent to take optimal actions (measured by total cumulative reward) through interaction with the environment and getting feedback signals. It is a class of optimization methods for solving sequential decision-making problems with a long-term objective in a stochastic environment. Thanks to the rapid advancement in deep learning research and computing power, the integration of deep neural networks and RL has generated explosive progress in solving complex large-scale decision problems (Silver and Hassabis 2016; Berner et al. 2019), attracting huge amount of renewed interests in the recent years. We are witnessing a similar trend in the ridesharing domain, where the demand and supply are highly stochastic and non-stationary, and the operational decisions are often sequential in nature and have strong spatiotemporal dependency. Simple greedy heuristics that only optimizes for immediate returns tend to produce short-sighted reactive policies, which do not align well with cumulative nature of the true performance measure of interest. The multi-step sequential

---

[1] The business model covered in this book is also often referred to as 'ride-sourcing' or 'on-demand ride services' (Wang and Yang 2019).

1
Z. (Tony). Qin et al., *Reinforcement Learning in the Ridesharing Marketplace*,
Synthesis Lectures on Learning, Networks, and Algorithms,
https://doi.org/10.1007/978-3-031-59640-7_1

nature of the decision-making (e.g., pricing, matching, repositioning) and the supply-demand stochasticity in the environment pose enormous challenges to traditional predictive and optimization methods, spanning such issues as forecast accuracy, decision-time computational complexity, and adaptability to real-time changes. RL, on the other hand, presents itself as an excellent promising approach to these ridesharing optimization problems. RL methods are often highly data-driven, making them more suitable to situations where it is hard to build accurate predictive models. They are forward-looking, and yet, they do not explicitly depend on forecasting. And, by design, RL-based policies are dynamic and often light in decision-time complexity.

There are excellent surveys on RL for intelligent transportation (Haydari and Yilmaz 2020; Yau et al. 2017), with in-depth coverage of traffic signals control and autonomous driving. Wang and Yang (2019) offers a broad review of ridesharing systems, whereas (Tong et al. 2020) surveys spatial crowdsourcing, which is a more general field than ridesharing. However, the ridesharing community is still in need of an in-depth technical book dedicated to the establishment of the knowledge base on RL for ridesharing, even though the field has attracted much attention and interest from the research communities for both RL and transportation just within the last few years. This book aims to fill that gap by surveying the literature of this domain published in top conferences and journals in transportation, data mining, and machine learning/AI, building upon (Qin et al. 2022), and deep-diving into a number of specific works covering the optimization of the main components of a ridesharing platform using RL. These are syntheses of our select works with collaborators and our practical experiences over the past eight years working in this field. We describe the research problems associated with the various aspects of a ridesharing system, review the existing RL approaches proposed to tackle each of them, describe the development of a few representative works, and discuss the challenges and opportunities.

This book is organized as follows: We lay out the ridesharing system architecture in Chap. 2 and define the scope of the topics covered in this book. In Chap. 3, we provide a concise review of the RL basics and the major algorithms adopted by the works discussed in this book. From Chaps. 4–8, we devote each chapter to one of the modules introduced in Chap. 2, in the order of their appearance in that chapter. We review in details the literature for the problems pertaining to each module in a ridesharing system, and then we dive in-depth on specific case studies that the book authors have worked on. In Chap. 9, we review two families of methods closely related to RL: approximate dynamic programming and model-predictive control. We survey the relevant data sets and environments in Chap. 10, and finally in Chap. 11, we discuss the major challenges and opportunities that we feel crucial in advancing RL for ridesharing.

Ridesharing 2

We first describe the architecture of a ridesharing system in this section, followed by explanation and clarification on the scopes of the problem associated with each module.

2.1 Mechanism

Traditional taxi hailing operates in a completely decentralized mode. Taxi drivers search for riders independently without any platform support, they pay a fixed annual fee for taxi medallion, and trips are priced with a fixed pricing scheme (also known as rate card). A ridesharing service, in contrast to taxi hailing, matches passengers with drivers of vehicles for hire using mobile apps. It is a platform-based service, with a single entry of demand. Drivers sign up and get onboard to the platform. Drivers are matched/dispatched to trip requests through a centralized system within the platform, which then charges commission fees on a per-request basis. Trips are priced by the platform usually in a more dynamic manner.

2.2 Architecture

In a typical mobile ridesharing system, there are five major modules: pricing, matching, repositioning, pooling, and routing. Figure 2.1 illustrates the process and decision modules. When a potential passenger submits a trip request, the *pricing* module offers a quote, which the passenger either accepts or rejects. Upon acceptance, the *matching* module attempts to assign the request to an available driver. Depending on driver pool availability, the request may have to wait in the system until a successful match. Pre-match cancellation may happen during this time. The assigned driver then travels to pick up the passenger, during which time post-match cancellation may also occur. The pick-up location is usually where the passenger is making the request or he/she specifies. In some cases, it could be a public designated area, e.g., outside an airport or train station. After the driver successfully transports the passenger to the destination, she receives the trip fare and becomes available again. The *repositioning* module guides idle vehicles to specific locations in anticipation of fulfilling more requests in the future. Following the reposition recommendations is usually on a voluntary basis unless it is an autonomous ridesharing setting. Hence, it is common that the platform offers incentives to drivers for completing the repositions. When each driver takes only one passenger request at a time, i.e. only one passenger shares the ride with the driver, this mode is more commonly called 'ride-hailing'. Ridesharing can refer to both ride-hailing and ride-pooling.[1] In the *ride-pooling* mode, multiple passengers with different trip requests can share one single vehicle, so the pricing, matching, repositioning, and routing problems are different from those for ride-hailing and require specific treatment, in particular, considering the passengers already on board. The *routing* module provides turn-by-turn guidance on the road network to drivers/vehicles either in service of a passenger request or performing a reposition. The goal is to guide the vehicle to its destination efficiently and safely.

2.3   Problem Scopes

First, we start from the pricing module. Since the trip fare is both the price that the passenger has to pay for the trip and the major factor for the income of the driver, pricing decisions influence both demand and supply distributions through price sensitivities of users, e.g., the use of surge pricing during peak hours. This is illustrated by the solid arrows pointing from the pricing module to orders and idle vehicles respectively in Fig. 2.1. The pricing problem in the ridesharing literature is in most cases dynamic pricing, which adjusts trip prices in real-time in view of the changing demand and supply. The pricing modules sits at the upstream position with respect to the other modules and is a macro-level lever to achieve supply-demand (SD) balance. Although technically, driver pay can be determined by a separate module from pricing and has its own implication on supply elasticity and driver behavior, this paper follows the common setting where driver pay is closely associated (approximately proportional) to the trip fare so that pricing has the dual effect on demand and supply.

The ridesharing matching problem (Yan et al. 2020; Özkan and Ward 2020; Qin et al. 2020a) may appear under different names in the literature, e.g., order dispatching (Qin et al. 2020a), trip-vehicle assignment (Bei and Zhang 2018), and on-demand taxi dispatching (Tong et al. 2020). It is an online bipartite matching problem where both supply and demand are dynamic, with the uncertainty coming from demand arrivals, travel times, and the entrance-exit behavior of the drivers. Matching can be done continuously in a streaming

---

[1] In this book, we do not cover topics on hitch, in which the driver is on his/her own trip with a specific destination.

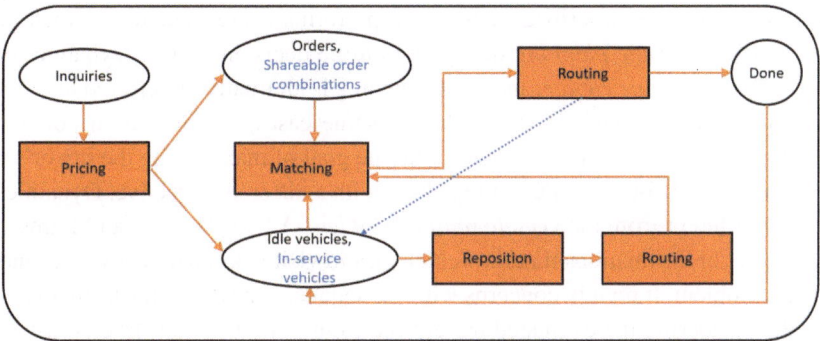

Fig. 2.1 The process flow of ridesharing operations. The solid orange rectangular boxes represent the modules described in this chapter. The blue text and arrow apply exclusively to ride-pooling to account for the fact that order combinations and in-service vehicles are also eligible to participate in matching

manner or at fixed review windows (i.e., batching). Sophisticated matching algorithms often leverage demand prediction in some form beyond the actual requests, e.g., the value function in RL. Online request matching is not entirely unique to ridesharing. Indeed, ridesharing matching falls into the family of more general dynamic matching problems for on-demand markets (Hu and Zhou 2022). A distinctive feature of the ridesharing problem is its spatiotemporal nature. A driver's eligibility to match and serve a trip request depends in part on her spatial proximity to the request. Trip requests generally take different amount of time to finish, and they change the spatial states of the drivers, affecting the supply distribution for future matching. The drivers and passengers generally exhibit asymmetric exit behaviors in that drivers usually stay in the system for an extended period of time, whereas passenger requests are lost after a much shorter waiting period in general.

Single-vehicle repositioning may refer to as taxi routing or passenger seeking in the literature. Taxi routing slightly differs in the setting from repositioning a rideshare vehicle in that a taxi typically has to be at a visual distance from the potential passenger to take the request whereas the matching radius of a mobile rideshare request is considerably longer, sometimes more than a mile. System-level vehicle repositioning, also known as driver dispatching, vehicle rebalancing/reallocation, or fleet management, aims to rebalance the global SD distributions by proactively dispatching idle vehicles to different locations. Repositioning and matching are similar to each other in that both relocate a vehicle to a different place as a consequence. In theory, one can treat repositioning as matching a vehicle to a virtual trip request, the destination of which is that of the reposition action, so that both matching and repositioning can be solved in a single problem instance. Typically in practice, these two problems are solved separately because they are separate system modules on most ridesharing platforms with different review intervals and objective metrics among other details.

The routing module described in Sect. 2.2 performs route guidance, which could be *dynamic routing* or *route planning* depending on the decision points. Dynamic routing is also called *dynamic route choice*, and route planning is alternatively referred to as the traffic assignment problem (Shou and Di 2020a). In some cases, the reposition policy directly provides link-level turn-by-turn guidance with the goal of maximizing the driver's income, thus covering the role of dynamic routing albeit with a different objective. Dynamic routing is generally different from the *vehicle routing problem* (VRP) (Dantzig and Ramser 1959). In VRP, the set of destinations that the vehicle has to visit is known in advance, and hence, it is a static problem. It mainly concerns with the sequence in which the destinations should be visited, considering the estimated travel time from point to point. In contrast, dynamic routing is associated with a road network, and the decision to make is which outgoing road (link) to follow at each intersection (node). The decision is adaptive to the changing traffic condition on the road network in real time. In the context of ride-pooling, there is another emerging problem in which the dynamic routing decisions with passenger(s) on board have to align with the overall objective of ride-pooling.

Some literature refers to the mode of multiple passengers sharing a ride as 'ridesharing'. In this paper, we use term '*ride-pooling*' (or 'carpool') to disambiguate the concept, as 'ridesharing' can refer to both single- and multiple-passenger rides. The seminal paper of Alonso-Mora et al. (2017a) shows that through ride-pooling optimization, the number of taxis required to meet the same trip demand can be significantly reduced with limited impact on passenger waiting times. In a pooling-enabled rideshare system, the matching, repositioning, and pricing modules all have to adapt to additional complexity. Compared to the regular ride-hailing problem, the one with ride-pooling has considerably more complexity due to the more dynamic nature of the problem and the additional constraints and multiple objectives that have to be considered. In this case, the set of available vehicles are augmented, including both idle vehicles and occupied ones not at capacity. It is non-trivial to determine the set of feasible actions (one or more passengers to be assigned to a vehicle) for matching. Every time a new passenger is to be added to a non-idle vehicle, the route has to be recalculated using a VRP solver to account for the additional pick-up and drop-off, the travel times for all the passengers already on board are updated, and the vehicle capacity, the waiting time and detour distance constraints are checked. In-service routing in the context of ride-pooling is discussed in Chap. 8.

We briefly review the RL basics and the major algorithms, especially those used in the works reviewed or described in this survey. For a complete reference, see, e.g., Sutton and Barto (2018).

3.1   Basics

RL is based on the Markov decision process (MDP) framework, where the agent (the decision-making entity) has a *state* s (e.g., the location of a vehicle) in the state space $\mathcal{S}$ and can perform an *action* a (e.g., to dispatch or idle) in the action space $\mathcal{A}$. The action is determined by the agent's policy, $\pi(s) : \mathcal{S} \to \mathcal{A}$. If the policy is stochastic, then $\pi(a|s)$ gives the probability of selecting a given s. After executing the action, the agent receives an immediate *reward* $R(s, a)$ from the environment, and its state changes according to the transition probabilities $P(\cdot|s, a)$. The process repeats until a terminal state or the end of the horizon is reached, giving a sequence of the triplets $(s_t, a_t, r_t)_{t=0}^{t=T}$, where t is the epoch index, T is the final epoch at the terminal state or the end of the horizon, and r_t is a sample of R. The objective of the MDP is to maximize the cumulative reward over the horizon. A key quantity to compute is the value function associated with π,

$$V^{\pi}(s) := E_{\pi}\left[\sum_{t=0}^{t=T} \gamma^t r_t \,\middle|\, s_0 = s\right],$$

which satisfies the Bellman equation,

$$V^\pi(s_t) = \sum_{a_t} \pi(a_t|s_t) \sum_{s_{t+1}, r_t} P(s_{t+1}, r_t|s_t, a_t)\Big(r_t(s_t, a_t) + \gamma V^\pi(s_{t+1})\Big). \tag{3.1}$$

Similarly, we have the action-value function,

$$Q^\pi(s, a) := E_\pi\left[\sum_{t=0}^{t=T} \gamma^t r_t \Big| s_0 = s, a_0 = a\right],$$

which conditions on both s and a. The optimal state- and action-values are denoted by V^* and Q^*, evaluated at the optimal policy π^*. The objective of an MDP is to find an optimal policy π^* that maximizes the long-term cumulative discounted reward, i.e., $\pi^* := \arg\max_\pi E_s[V^\pi(s)]$.

3.2 Algorithms

Given P and R, which specifies the MDP, and π, we can compute V^π by iteratively applying the Bellman equation (3.1),

$$V^\pi(s) \leftarrow \sum_a \pi(a|s) \sum_{s', r} P(s', r|s, a)\Big(r + \gamma V^\pi(s')\Big). \tag{3.2}$$

This is called *policy evaluation*. Using policy evaluation as a sub-routine, we can again iteratively improve the policy through *policy iterations*, which generate a new policy at each outer iteration by acting greedily with respect to V^π,

$$a^* \leftarrow \arg\max_a \sum_{s', r} P(s', r|s, a)\Big(r + \gamma V^\pi(s')\Big). \tag{3.3}$$

As a special instance, we can collapse the inner policy evaluation loop to a single iteration and compute $V^*(s), \forall s \in \mathcal{S}$, by the *value iterations*,

$$V(s) \leftarrow \max_a \sum_{s', r} P(s', r|s, a)\Big(r + \gamma V(s')\Big). \tag{3.4}$$

If we estimate (P, R) from data, we have a basic model-based RL method. A model-free method learns the value function and optimizes the policy directly from data without learning and constructing a model of the environment. A common example is *temporal-difference (TD) learning* (Sutton 1988), which iteratively updates V^π by TD-errors using π-generated trajectory samples and bootstrapping,

$$V^\pi(s) \leftarrow V^\pi(s) + \alpha\left(r + \gamma V^\pi(s') - V^\pi(s)\right), \tag{3.5}$$

where s' is the next state in the trajectory after s, α is the step size (or learning rate), and the term that it scales is the TD error. If learning the optimal action-values for control is the goal, we similarly have *Q-learning* (Watkins and Dayan 1992), which updates the action-value function $Q(s, a)$ to approximate $Q^*(s, a)$ by

$$Q(s, a) \leftarrow Q(s, a) + \alpha\left(r + \gamma \max_{a'} Q(s', a') - Q(s, a)\right). \tag{3.6}$$

Q-learning is an *off-policy* algorithm, where the behavior policy, which collects the experience data and typically involves exploration, is different from the target policy that we are trying to learn and in this case, is the optimal policy. The *on-policy* counterpart of Q-learning is SARSA, which basically generalizes TD-learning (3.5) to the action-value function associated with the behavior policy π (same as the target policy):

$$Q^\pi(s, a) \leftarrow Q^\pi(s, a) + \alpha\left(r + \gamma Q^\pi(s', a') - Q^\pi(s, a)\right), \tag{3.7}$$

where a' is the action executed by the agent at state s' in the experience data. The *deep Q-network* (DQN) (Mnih et al. 2015) approximates $Q(s, a)$ by a neural network Q_w parametrized by w along with a few heuristic techniques like experience replay (Lin 1992) and a target network to improve training stability. These techniques are critical to the successful of DQN in playing Atari games and many other applications, due to the deadly triad issue (Sutton and Barto 2018) of reinforcement learning when one tries to combine bootstrapping (i.e., TD-learning and Q-learning), off-policy training (i.e., Q-learning), and function approximations (i.e., neural networks), which may lead to instability and divergence. The algorithms introduced so far are all value-based methods, which focus on learning the value function, and the policy is derived from the learned value function by, e.g., $\arg\max_a Q(s, a)$. Neural network-based value function approximation is important to ridesharing applications because the state is often high dimensional with the incorporation of SD contextual features. Tabular methods suffer from the curse of dimensionality and are not tractable in this case.

A policy-based method directly learns π (which is also called the *actor* and parametrized by θ) by performing stochastic gradient descent. The central step is computing the policy gradient (PG), the gradient of the cumulative reward $J(\theta)$ with respect to the policy parameters θ,

$$\nabla_\theta J(\theta) = \sum_s \mu(s) \sum_a Q^\pi(s, a) \nabla_\theta \pi(a|s, \theta), \tag{3.8}$$

where μ is the on-policy distribution under π. A more common (equivalent) form of the PG (3.8) is

$$\sum_s \mu(s) \sum_a \pi(a|s, \theta) Q^\pi(s, a) \nabla_\theta \log \pi(a|s, \theta), \tag{3.9}$$

which most of the policy-based methods are based on. REINFORCE (Williams 1992) is a classical PG method, which uses Monte Carlo (MC) rollout to obtains the sample-based update

$$\theta_{t+1} \leftarrow \theta_t + \alpha G_t \nabla_\theta \log \pi(a|s, \theta), \qquad (3.10)$$

where G_t is an MC approximation of Q^π. As in a value-based method, we can also use function approximation for the action values. The function approximator (e.g., a neural network) Q_w is called the *critic*, and the resulting algorithm is an *actor-critic* (AC) method. It is well-recognized that the 'baseline' version of (3.8),

$$\sum_s \mu(s) \sum_a \pi(a|s, \theta) \left(Q^\pi(s, a) - b(s) \right) \nabla_\theta \log \pi(a|s, \theta), \qquad (3.11)$$

where $b(s)$ is an action-independent baseline, reduces the variance in the sample gradient and helps speed up learning. Since a natural choice of such a baseline is the state value, the critic often learns the advantage function $Q(s, a) - V(s)$, and the method is called Advantage Actor Critic (A2C). The evaluation of an action is based on how good it can be with respect to the average over all actions, the benefit of which is to reduce the high variance in the actor and to stabilize the model. Mnih et al. (2016) extend A2C to an asynchronous version (A3C) where independent agents interact with their own copy of the environment and update their model parameters with the master copy asynchronously. This architecture enables much more efficient utilization of the CPU cores through parallel threads and hence accelerates the training. The proximal policy optimization (PPO) (Schulman et al. 2017) optimizes a clipping surrogate objective with respect to the advantage to promote conservative updates to π and is a popular choice of training algorithm for RL problems where policy-based methods are suitable (e.g., with continuous actions).

An MDP can be extended to a Markov game involving multiple agents to form the basis for multi-agent RL (MARL) (Buşoniu et al. 2010). Many MARL algorithms, e.g., Yang et al. (2018), Lowe et al. (2017b) focus on agent communication and coordination, in view of the intractable action space in MARL. Interested readers can refer to Oroojlooy and Hajinezhad (2022) for a review of cooperative multi-agent deep RL methods.

Recently, an emerging stream of literature has focused on marrying the merits of causal modeling and inference with reinforcement learning to tackle the long-standing challenges in generalization, sample efficiency, and interpretability of RL. By incorporating causality into the learning process, causal RL aims to enhance RL algorithms in better understanding of the environment. We review and discuss relevant works leveraging causal RL for ridesharing problems in Chap. 9. For a more comprehensive treatment of the causal RL literature, We refer the reader to the recent surveys (Deng et al. 2023; Zeng et al. 2023).

Pricing & Incentives

4

4.1 Dynamic Pricing

RL-based approaches have been developed for dynamic pricing in one-sided retail markets (Raju et al. 2003; Bertsimas and Perakis 2006), where pricing changes only the demand pattern per customers' price elasticity. The ridesharing marketplace, however, is more complex due to its two-sided nature and spatiotemmporal dimensions. In this case, pricing is also a lever to change the supply (driver) distribution if price changes are broadcast to the drivers. Chen et al. (2021a) describe examples of such elasticity functions for both demand and supply for their simulation environment.

The challenges in dynamic pricing for ridesharing lie in both its exogeneity and endogeneity. Dynamic pricing on trip inquiries changes the subsequent distribution of the submitted requests through passenger price elasticity. The requests distribution, in turn, influences future supply distribution as drivers fulfill those requests. On the other hand, the trip fares influence the demand for ridesharing services at given locations, and these changes will affect the pool of waiting passengers, which further affects the passengers' expected waiting times. Again, it will influence the demand either through cancellation of the current requests or the conversion of future trip inquiries. Because of its close ties to SD distributions, dynamic pricing is often jointly optimized with order matching or vehicle repositioning. Within the (non-RL) operations research literature, dynamic pricing for ridesharing has already been studied and analyzed in conjunction with matching (Yan et al. 2020; Özkan and Ward 2020) and from the spatiotemporal perspective (Ma et al. 2020; Bimpikis et al. 2019; Hu et al. 2022), covering optimality and equilibrium analyses.

The complex interaction between pricing and the SD makes it hard to explicitly develop mathematical models that adapt well to dynamic and stochastic environments, and RL comes in as a promising direction to address these challenges by considering endogeneity and exogeneity as part of the environment dynamics.

© The Author(s), under exclusive license to Springer Nature Switzerland AG 2025   11
Z. (Tony). Qin et al., *Reinforcement Learning in the Ridesharing Marketplace*,
Synthesis Lectures on Learning, Networks, and Algorithms,
https://doi.org/10.1007/978-3-031-59640-7_4

Table B.1 summarizes the reviewed works on RL for dynamic pricing in ridesharing. As one of the early RL works, Wu et al. (2016) consider a simplified ridesharing environment which captures only the two-sidedness of the market but not the spatiotemporal dimensions. The state of the MDP is the current price plus SD information. The action is to set a price, and the reward is the generated profit. A Q-learning agent is trained in a simple simulator, and empirical advantage in the total profit is demonstrated against other heuristic approaches. More recent works leverage the spatiotemporal nature of the pricing actions and take into account the spatiotemporal long-term values in the pricing decisions. Chen et al. (2019a) integrate contextual bandits and the spatiotemporal value network developed in Tang et al. (2019) for matching to jointly optimize pricing and matching decisions. In particular, the pricing actions are the discretized price percentage changes and are selected by a contextual bandits algorithm, where the long-term values learned by the value network are incorporated into the bandit rewards. In Turan et al. (2020), the RL agent determines both the price for each origin-destination (OD) pairs and the reposition/charging decisions for each electric vehicle in the fleet. The state contains global information such as the electricity price in each zone, the passenger queue length for OD pair, and the number of vehicles in each zone and their energy levels. The reward accounts for trip revenue, penalty for the queues, and operational cost for charging and repositioning. Due to the multi-dimensional continuous action space, PPO is used to train the agent in a simulator. Song et al. (2020) perform a case study of ridesharing in Seoul. They use a tabular Q-learning agent to determine spatiotemporal pricing, and extensive simulations are performed to analyze the impact of surge pricing on alleviating the problem of marginalized zones (areas where it is consistently hard to get a taxi) and on improving spatial equity. Chen et al. (2021a) adopt PPO to optimize the spatiotemporal pricing decisions for each hexagonal cell in terms of the per-km rate for the excess mileage beyond a base trip distance and the per-km rate for driver wage, for the objective of maximizing profits (revenue minus wage). The agent is modeled as a global decision-maker with state information of the numbers of open requests, vacant vehicles, occupied vehicles in each grid cell at time t and historical demand at time $t-1$. Unlike the works above that focus on the pricing decisions, Mazumdar et al. (2017) study from a different perspective of the pricing problem. The proposed risk-sensitive inverse RL method (Ng et al. 2000) recovers the policies of different types of passengers (risk-averse, risk-neutral, and risk-seeking) in view of surge pricing. The policy determines whether the passenger should wait or take the current ride.

As discussed in Sect. 2.3, under the setting where the driver pay is associated with the trip fare, the dynamic pricing policy also affects supply elasticity, i.e., drivers' decisions on participation in a given marketplace, working hours, and in some cases, the probability of accepting a given assignment, depending on the rules of the particular ridesharing platform (Chen and Sheldon 2016; Sun et al. 2019; Angrist et al. 2021). Although not yet being widely considered in RL approaches to pricing, supply elasticity is an important piece of system state information that has significant implication to the sequence of pricing decisions. The problem becomes more complex when driver pay deviates more from a constant fraction of

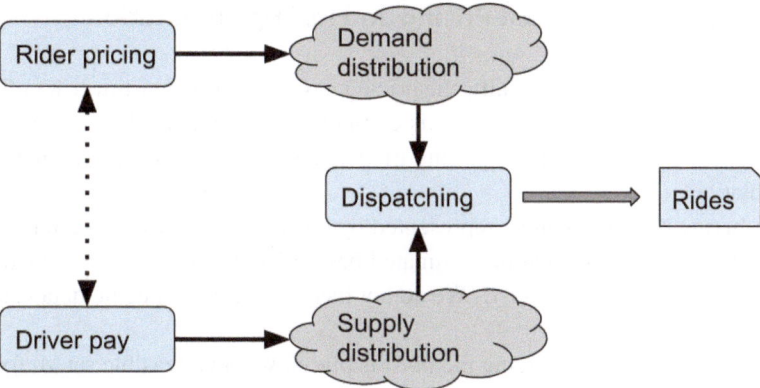

Fig. 4.1 Interaction of rider pricing and driver pay

the trip fare. In this case, the dynamic pricing and driver pay policies have to be coordinated to make sure that the platform is overall profitable and sustainable. Figure 4.1 illustrates the interaction between dynamic pricing and driver pay and how they influence the inputs to dispatching, which outputs matched rides.

4.2   Case Study: Dispatch-Aware Dynamic Pricing Optimization

A pricing strategy consists of two components, (1) a base price which is a fixed price determined by the travel distance and travel time, etc., and (2) a pricing factor which is a multiplication over the base price. In this section, we describe a dispatch-aware dynamic pricing algorithm developed in Chen et al. (2019a) for adjusting the pricing factor. (In this setting, the base price is an external input.) One key idea of our approach is that, instead of focusing on the immediate effect of a pricing strategy, we also consider its future effects. Intuitively, we encourage, with a reduced price, the conversion to orders of requests from an over-supplied area to an under-supplied area. Later, we will show how to quantitatively characterize this property. After a driver is assigned to the passenger and drives the passenger to a hot area, the driver is more likely to be able to fulfill another order immediately. This mitigates the supply-demand imbalance, while improving the operational efficiency of the platform. As we can see, the future effect of a request pricing strategy is reflected in the repositioning of a driver, from its original position at the current time to its destination at a future time. Thus, the future effect of repositioning the driver is highly dependent on the dispatch strategies, making it sub-optimal to optimize pricing without considering dispatching.

4.2.1 Distributed Request Pricing and Its Implementation

As ride requests emerge dynamically, the pricing has to occur in real-time (in milliseconds). Global optimization of pricing strategy is computationally infeasible, pushing us towards an individually optimized request pricing model, ensuring both scalability and distributed implementation.

A ridesharing request i can be represented by a vector of contextual features, including time of request, origin, destination, estimated base price, trip distance, and historical conversion rate, signified by $x_i = \langle x_{ij} \rangle$. We do not incorporate user-specific details to keep the framework generalized.

Over the base price p_i, a pricing factor a_i is placed within a feasible set $\mathcal{A}$, for instance, $\mathcal{A} = \{0.85, 0.9, \ldots, 1.15\}$. As one might expect, the conversion rate function $f(x_i, a_i)$, indicating the likelihood of a ride request transpiring into a completed order, is anticipated to decrease with an increasing pricing factor.

The immediate reward for employing a pricing strategy a_i for request i, with an order being dispatched and completed, is expressed as:

$$r(x_i, a_i) = f(x_i, a_i)(p_i a_i - p_i \beta),$$

where β represents the ride-hailing platform's revenue share.

However, mere consideration of immediate rewards fails to account for the subtle nuances involved in a driver's potential to earn future income. This ushers in the necessity for a value function that encapsulates both immediate and future rewards.

4.2.2 Value Function in the Realm of Pricing and Dispatching

When integrating dispatch mechanics, the immediate reward, essentially the driver income, evolves into:

$$r_\pi(x_i, a_i) = \sum_{j \in \mathcal{J}_i} f(x_i, a_i) b_{ji}(p_i a_i - p_i \beta).$$

Here, $\mathcal{J}_i$ signifies the set of drivers relevant to request i, and b_{ji} implies if driver j is assigned to request i.

The key to our integrated framework lies in the future reward component, which factors in the repositioning of the driver after ride completion. With a driver's spatio-temporal value signified by $V_\pi(l, t)$ and considering a discount factor γ, the future reward aligns as:

$$R_\pi(x_i, a_i) = \gamma \sum_{j \in \mathcal{J}_i} f(x_i, a_i) b_{ji}(V_\pi(l_i', t_j + T_i) - V_\pi(l_j, t_j)),$$

which essentially captures the change in a driver's value from origin to destination. The total expected reward then stitches together the immediate and future rewards, adding depth to our pricing decisions. In practice, this can be estimated by

$$\hat{u}_\pi(x_i, a_i) = \sum_{i \in \mathcal{J}_i} f(x_i, a_i) b_{ji} [p_i a_i - p_i \beta$$
$$+ \gamma(\hat{V}_\pi(l'_i, t_j + T_i; \phi) - \hat{V}_\pi(l_j, t_j; \phi))] \tag{4.1}$$

where $\hat{V}_\pi(l_j, t_j; \phi)$ is an approximation of the long term spatio-temporal value of a driver with parameter ϕ (e.g., a neural network).

Despite such mathematically clear expositions, learning the precise spatio-temporal value function of drivers and request conversion functions pose significant challenges, more so when interdependencies amongst pricing and dispatch policies are considered. This complexity is accentuated by the latency requirements of pricing decisions and the vast state and action spaces we tend to work with.

To circumvent these hurdles, we introduce the InBEDE algorithm.

4.2.3   InBEDE: A Paradigm Shift in Pricing Optimization

InBEDE stands unique, amalgamating contextual bandit approaches for pricing with TD learning to estimate future pricing effects, thus bridging the gap between request pricing and order dispatch. The beauty of this approach is in its recursive boosting. The contextual bandit model and driver spatio-temporal value function mutually enhance each other's accuracy with each iteration.

Here is how the methodology unfolds at a high level: 1. Split the pricing optimization task into *semi-contextual bandit* problems, where each ride request emerges as a trial, and a set of pricing factors form arms of the bandit algorithm. 2. Estimate the expected payoff for each arm using LinUCB or similar algorithms, considering not only the immediate reward but also an estimated future reward based on the current policy. This is where the algorithm differs from a traditional contextual bandit method, where the payoff function is based on immediate rewards. 3. Employ TD learning to estimate the future impact of pricing strategies, using estimated spatiotemporal driver values to assess the expected long-term value of pricing decisions. In Sect. 5.8, we will describe in details how to construct and learn a spatiotemporal driver value network for this purpose. Figure 4.2 illustrates this algorithmic framework, and we defer the detailed algorithm statement to Algorithm A.1 in the appendix.

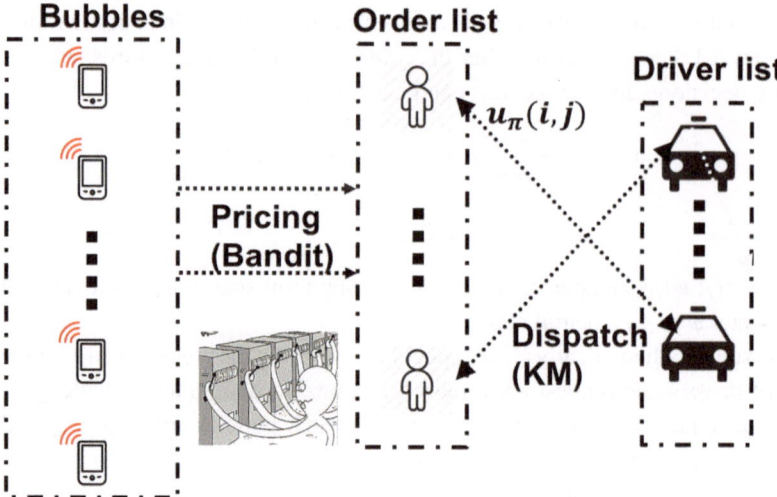

Fig. 4.2 An integrated approach for request pricing and order dispatch. Dispatch edge weights are updated every day. Bandit policy parameters are updated per dispatch step. *Source* Chen et al. (2019a)

4.2.4 Evaluation

In our empirical analysis against benchmarks, InBEDE exhibits superior performance in terms of total reward across varying price elasticity scenarios (see Fig. 4.3 and more details in Chen et al. (2019a)). Evidently, with an increase in price sensitivity (or elasticity), there's a discernible decrease in total reward as demand becomes more price-elastic. In practice, this is consistent with the economic principle of diminishing returns with heightened prices.

Notably, InBEDE outperforms traditional methods like fixed-value-network dispatch frameworks, thanks to its employment of iterative learning. This iterative method allows

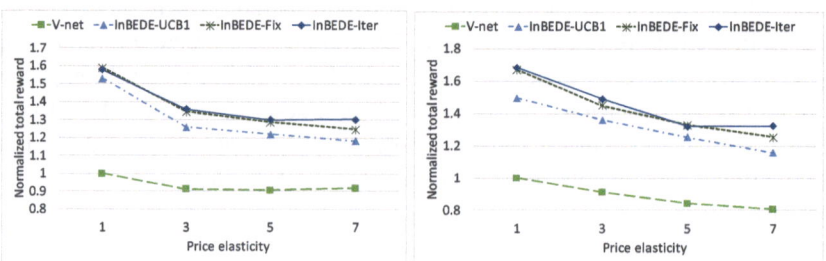

Fig. 4.3 Total reward of different methods on a contextual CR function setting. The x-axis denotes the price elasticity k, and the y-axis denotes the total reward, which is normalized by dividing the total reward of the V-net method with $k = 1$. Left: City A; Right: City B

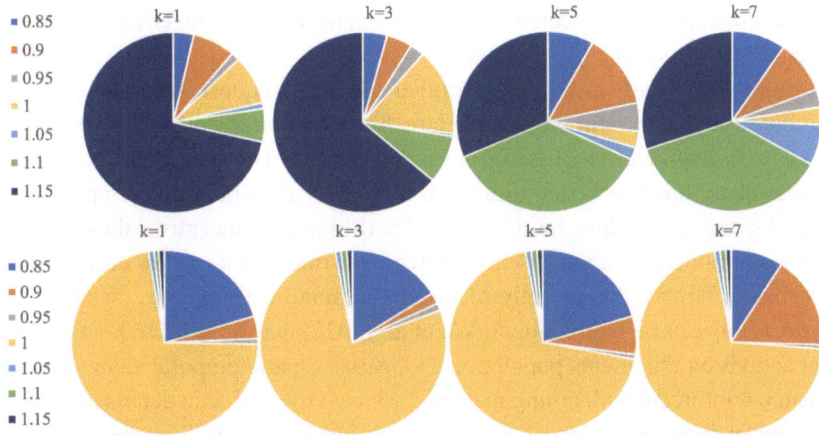

Fig. 4.4 Statistics of the price selected by InBEDE-Iter in the contextual CR function scenario, with varying price elasticity values from $k = 1$ to 7. Top: City A; Bottom: City B

for seamless adaptations in the face of real-world complexities, making it more representative of actual user behavior and marketplace dynamics.

Additionally, InBEDE uncovers an intriguing observation: it is potentially more profitable to quote higher fares per request, contrary to the common pricing strategy of most ride-hailing platforms. This runs counter to intuitions that would otherwise lower prices for consumer attraction, indicating a potential underpricing phenomenon in the industry focused mainly on customer satisfaction and competition (Fig. 4.4, City A).

InBEDE posits a powerful dynamic pricing algorithm that beautifully synchronizes with dispatch strategies, ensuring a long-term valuation of rideshare requests. Its iterative learning component is indispensable, allowing a real-time adjustment that acknowledges both the immediate and longer-term implications of pricing decisions.

4.3 Incentives Optimization

Incentives optimization is broadly used in a diverse array of domains beyond ridesharing, such as retail (traditional and online) and delivery. General targeting algorithms are extensively studied by the recommendation and causal inference communities. Here, we focus on the particular aspects related to ridesharing—the spatiotemporal nature of the incentives and the temporal dependency in their sequential decisions. Rideshare incentives generally divides into two categories: rider incentives and driver incentives, targeted to the key participant populations in the two-sided marketplace. Both rider and driver incentives can be further classified into real-time and offline/batch incentives, designed for different purposes. Despite the various forms, rideshare incentives are all evaluated by spending efficiency (incremental return per dollar spent, on a given budget) for success.

Real-time rider incentives are often applied at the pre-request stage to ride intents, typically in the form of discount off the trip fare. The decisions to make are what percentage of the fare to discount (zero included) for each ride intent. The goal is to dynamically shape the demand distribution to better align with the supply distribution over an extended spatiotemporal space. Algorithms for this class of real-time rider incentives thus solve a similar problem to dynamic pricing focused on price adjustment (e.g., Chen et al. 2019a), with the important difference that incentive decisions are unidirectional (i.e., discounts) whereas pricing decisions can be bidirectional. Wu et al. (2022) builds an MDP for the pre-request discount actions on the intents population and learns a spatiotemporal value function using offline batch reinforcement learning methods, which we describe in details in Sect. 5.8. The actual intent-level discounts are determined by solving a knapsack-like integer programming problem to ensure budget feasibility.

Batch incentives for riders are mainly to conduct the customer life-cycle management (AboElHamd et al. 2020; Theocharous et al. 2015). Newly registered riders, for example, are often given coupons to incentivize continued usage of the platform and prevent early-stage churning. Coupons could also be targeted to deeply churned riders (who have no activity with the platform for an extended period of time) to revitalize those users. We can see RL for rider batch incentives as a particular implementation of customer life-time value (LTV) (Theocharous et al. 2015) optimization. Representative works on this type of sequential targeting problems are Pednault et al. (2002), Abe et al. (2004) for direct mail campaigns for cross-channel marketing using batch RL and Abe et al. (2010) for debt collection optimization with cost constraints. Zhao et al. (2018) proposes a method of deep reinforcement learning in the recommendation system, through the interactions with users and receiving reinforcements of these items from users' feedback, rather than considering the recommendation procedure as a static process. Hao et al. (2020) proposes a dynamic knapsack method to optimize multi-channel sequential advertising to maximize the long-term cumulative revenue over a period of time under a budget constraint.

Real-time driver incentives are for repositioning and supply shaping (e.g., Ong et al. 2021). They are often in the form of a bonus conditional on the driver being present at a specific location within a prescribed time frame, with the goal of improving the alignment of the supply and demand distributions. We refer the reader to Chap. 6 for details about driver repositioning.

Offline batch driver incentives are typically multi-tiered target-based incentives (Huang et al. 2022) (also known as 'ride streaks'[1]), which offer bonuses to the participating drivers if they complete a specific number of rides within a given time frame (day or hours). The bonus structure can be laddered with different bonus amounts for different stackable target levels. The bonus amount and target levels are all configurable parameters. Liu et al. (2023b), Liu et al. (2023a) investigate in-depth about the effects and implications of these target-based

---

[1] https://help.lyft.com/hc/e/driver/articles/115015748908-Streak-Bonus.

incentives on drivers' labor supply behavior and supply management. Shang et al. (2019) develops a generative algorithm for automatically constructing the simulation environment for batch driver incentives from historical data, within which one can employ RL algorithms for policy optimization. Shang et al. (2021), Huang et al. (2022) further investigate uplift modeling and causal inference for the reward function required in RL-based optimization methods for this class of incentives. Alternatively, we can leverage offline data in a more direct manner through offline learning and evaluation. Chen et al. (2021b), Chen et al. (2023) propose offline RL methods motivated by the batch driver incentives problem and further investigate sim2real transfer and adaptation to close the gap in applying such methods in real-world settings. Similarly motivated by the system-level driver and rider incentives problems, Shi et al. (2023a) develop a multi-agent RL approach (with spatial units as agents) for off-policy evaluation (OPE) on batch data.

In addition to the incentives for regular ride-hailing, there are incentives specifically designed for ride-pooling. These incentives are typically route-based to improve the share-ability of the requests and the match efficiency (see Chap. 8) of the carpool service.

Although individual forms of the incentives discussed above have been studied within an RL context, little investigation has been done on the interaction between the rider and driver incentives. Nevertheless, the coordination between those two types of levers is important since the coherence between supply and demand is key to an efficient rideshare marketplace.

4.4 Case Study: Learning-Based Environment Construction for Driver Incentives

Decision-making for offline driver incentives primarily involves driver targeting and incentive configuration. Specifically, the platform targets the appropriate incentive programs to a subset of the drivers every day and adapts the program configuration according to the drivers' responses. This is a typical sequential recommendation task and can be naturally solved by reinforcement learning. However, the driver behaviors are influenced by not only the recommended incentive programs but also other unobservable factors, such as special events. It is thus essential to take into account the potential influence of latent factors when optimizing incentives program recommendation. Traditional reinforcement learning approaches are applied to these problems without exploring the impact of hidden variables, which would consequently degrade the learning performance. Shang et al. (2019), Shang et al. (2021) develop a learning-based approach to construct a generative model for driver behavior with respect to the incentives policy and latent factors based on historical data and subsequently trains an RL agent through interacting with the generated environment to optimize the incentives design for system-level metrics. Such simulator-based reinforcement learning method can be very efficient without any interaction cost and risk with the real-world environment. In this case study, we describe this novel *partially-observed multi-agent environment estimation* (POMEE) method.

4.4.1   Background

Imitation Learning

Imitation learning, a technique that draws inspiration from demonstrated expert behavior, has found practical utility in numerous applications (Ross et al. 2011). The two main strands of imitation learning are behavioral cloning and inverse reinforcement learning (Pomerleau 1991; Russell 1998). Commonly, imitation learning can be stated as the training of policy π to minimize the loss function $l(s, \pi(s))$, under the discounted state distribution of the expert policy. This can be denoted as the following equation:

$$\pi = \arg\min_{\pi} \mathbb{E}_{s \sim P_{\pi_e}}[l(s, \pi(s))], \qquad (4.2)$$

where $P_{\pi_e}(s) = (1 - \gamma)\Sigma_{t=0}^{T}\gamma^t p(s_t)$.

Generative Adversarial Imitation Learning

Generative Adversarial Imitation Learning (GAIL) is a popular imitation learning method, mitigating some of the traditional imitation learning drawbacks, like the instability of behavioral cloning and complexity of inverse reinforcement learning (Ho and Ermon 2016). GAIL uses the Generative Adversarial Network (GAN) framework to learn a policy with guidance from a reward function, given expert demonstrations as real samples. The objective function used in GAIL is comparable to the GAN objective function and can be denoted by the following equation:

$$\arg\min_{\pi} \arg\max_{D \in (0,1)} \mathbb{E}_{\pi}[\log D(s, a)] + \mathbb{E}_{\pi_E}[\log(1 - D(s, a))] - \lambda H(\pi), \qquad (4.3)$$

where $H(\pi) \triangleq \mathbb{E}_{\pi}[-\log \pi(a|s)]$ is the entropy of π.

In this framework, the agent obtains the ability to execute the policy in the environment, update it with policy gradient methods and optimize it to maximize the similarity between policy-derived trajectories and expert ones (Schulman et al. 2015). The policy π can be updated to minimize the loss function:

$$l(s, \pi(s)) = \mathbb{E}_{\pi}[\log D(s, a)] - \lambda H(\pi) \cong \mathbb{E}_{\tau_i}[\log \pi(a|s)Q(s, a)] - \lambda H(\pi), \qquad (4.4)$$

where $Q(s, a) = \mathbb{E}_{\tau_i}[\log(D(s, a))|s_0 = s, a_0 = a]$ is the state-action value function.

GAIL has proven to achieve similar results as IRL, but more efficiently (Finn et al. 2016). A multi-agent extension to GAIL has also shown promise in reconstructing environments (Shi et al. 2019b).

4.4.2   Partially-Observed Environment Estimation Through Multi-agent Imitation Learning

The hidden variables influencing driver behavior are modeled as an unobservable agent in our multi-agent environment. We treat this unobservable agent as a hidden policy while the observable drivers and platforms form observable agents. Interactions between agents take the form of simulated trajectories, beginning with the agent A taking a decision $a_A = \pi_a(o_A)$, based on its observation o_A. The hidden agent H's observation and action are then reliant on this initial decision by agent A. The hidden agent's decision in turn influences the response of the environment agent E.

Our goal is to use observable interactions to imitate and characterize the policies π_a and π_e while inferring the hidden policy π_h. We do this by training a policy π_a and a joint policy $\pi_{he} = \pi_e \circ \pi_h$ to minimize the loss functions

$$l(o_A, \pi_a(o_A)) = \mathbb{E}_{\pi_a}[\log D_a(o_A, a_A)] - \lambda H(\pi_a)$$
$$\cong \mathbb{E}_{\tau_i}[\log \pi_a(a_A|o_A)Q(o_A, a_A)] - \lambda H(\pi_a), \qquad (4.5)$$

and

$$l((o_A, a_A), \pi_{he}((o_A, a_A))) = \mathbb{E}_{\pi_h, \pi_e}[\log D_{he}((o_A, a_A), a_E)] - \lambda \Sigma_{\pi \in \{\pi_h, \pi_e\}} H(\pi)$$
$$\cong \mathbb{E}_{\tau_i}[\log \pi_{he}(a_E|o_A, a_A)Q(o_A, a_A, a_E)] - \lambda \Sigma_{\pi \in \{\pi_h, \pi_e\}} H(\pi), \qquad (4.6)$$

respectively.

With these results, we propose POMEE to accomplish the objective of capturing the hidden aspects of our environment.

Partially-Observed Environment Model

In our POMEE model, we deal with two observable agents, namely a policy agent A and an environment E, as well as one unobservable agent H representing hidden variables. Using the results from the breakdown of the objective function, we merge the hidden policy π_h and observable policy π_e into a joint policy, denoted as $\pi_{he} = \pi_e \circ \pi_h$.

The policies between the generator are updated alternately during the training process: initially, the joint policy π_{he} is updated using an imitation reward given by the discriminator, followed by the update of the policy π_a—thereby implicitly inferring the hidden policy— using a model such as TRPO to stabilize the training process (Schulman et al. 2015).

Compatible Discriminator for Multiple Tasks

A compatible discriminator is necessary for multi-task learning from the generator, as the generator is tasked with simulating and learning different reward functions for policies π_a and π_{he} respectively. The discriminator makes use of two classifier tasks to evaluate the state-action pairs of both the policy π_a and the joint policy π_{he}. The loss function of each task is defined as

$$E_{\tau_{sim}}[\log(D_\sigma(o_A, a_A, a_E))] + E_{\tau_{real}}[\log(1 - D_\sigma(o_A, a_A, a_E))] \qquad (4.7)$$

for π_{he} , and

$$E_{\tau_{sim}}[\log(D_\sigma(o_A, a_A, \mathbf{0}))] + E_{\tau_{real}}[\log(1 - D_\sigma(o_A, a_A, \mathbf{0}))] \qquad (4.8)$$

for policy π_a.

The output of the discriminator signifies the likelihood that specific pair data sources derive from real rather than generated data. The discriminator's methodology involves training via labeling the real state-action pairs as '1' and generated pairs as '0', before being used as a reward function for the policies in the simulation process. The reward function for policy π_{he} can be written as:

$$r^{he} = -\log(1 - D(o_A, \ a_A, \ a_E)), \qquad (4.9)$$

and the reward function for policy π_a is

$$r^a = -\log(1 - D(o_A, \ a_A, \ \mathbf{0})). \qquad (4.10)$$

Figure 4.5 summarizes POMEE in a schematic illustration, and we defer the precise algorithm statement to Algorithm A.2 in the appendix.

Implementation: Partially-Observed Multi-agent Environment Estimation

With an understanding of the partially-observed environment model and compatible discriminator, we develop the POMEE approach to simulate the hidden effects on drivers from observed interactions. The partially-observed environment model simulates interactions between the platform and drivers using policies π_a and π_{he}. Following this simulation process, these policies are updated using an imitation reward from the discriminator.

4.4.3   Application: Driver Program Recommendation

The POMEE model is applied to construct a virtual environment that simulates the driver program recommendation using historical data (see Fig. 4.6 for a schematic illustration). The implementation considers three agents, representing the driver policy π_d, platform policy π_p, and hidden policy π_h. The platform policy serves as the mutual environment for the driver and hidden agents (Shang et al. 2021).

We define the observation and action of each agent policy as follows: The observation for platform policy π_p comprises driver's past behaviour and static characteristics, and the recommended program makes up the action. For the hidden policy π_h, the observation is the

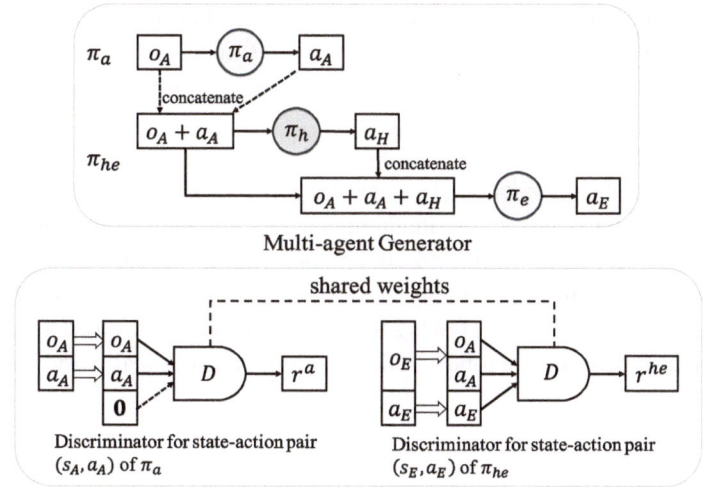

Multi-agent Generator

Compatible Discriminator

Fig. 4.5 The generator and the discriminator in POMEE. The multi-agent interactive environment plays a role of generator, and can generate simulation interaction data. The discriminator is designed to be compatible for classify the state-action pairs of both the policy π_a and the joint policy π_{he}. *Source* Shang et al. (2021)

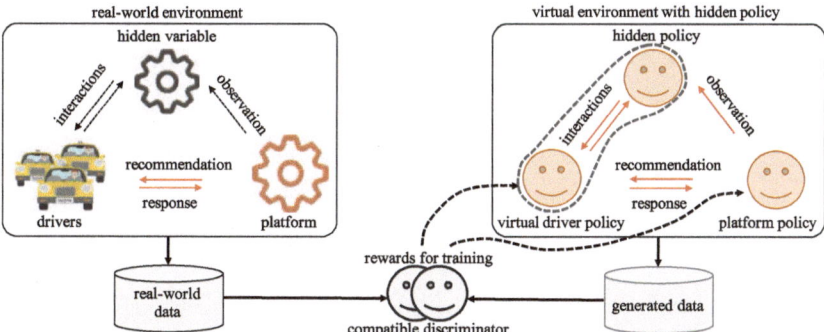

Fig. 4.6 POMEE framework applied in the driver program recommendation. While real-world data only collects the interactions between the drivers and the ridesharing platform, the virtual environment contains three policies simulating the drivers, the platform, and the hidden variable. *Source* Shang et al. (2021)

concatenation of the platform's observation and its action, while its action format mimics that of the platform policy. The driver policy π_d's observation includes the platform's observation and both the platform and hidden agent's actions, and the response to the platform's program constitutes the action.

We train π_d and π_h using POMEE and reconstruct the partially-observed environment of driver program recommendation. Through this environment, we can use reinforcement learning to optimize the policy π_p by simulating interactions with the environment. In Shang et al. (2021) (also see Huang et al. (2022)), an uplift model is further developed to embed in the virtual environment for modeling the reward, which has a good causal relationship with the recommended programs. Owing to the simulated hidden variables' presence, the reinforcement agent can learn a robust policy resulting in better real-world performance.

Online Matching (Dispatching)

5

The rideshare matching problem and its generalized forms have been investigated extensively in the field of operations research (see e.g., Özkan and Ward 2020; Hu and Zhou 2022; Lowalekar et al. 2018 and the references therein). Typically, both the open trip requests and available drivers are batched within time windows of fixed length as they arrive at the system, and they are matched at predefined discrete review times. See Fig. 5.1 for an illustration. Hence, ridesharing matching is an online stochastic problem (Qin et al. 2020a). (See Sect. 5.1 for a mathematical formulation.)

Outside the RL literature, Lowalekar et al. (2018) approach the problem through stochastic optimization and use Bender's decomposition to solve it efficiently. To account for the temporal dependency of the decisions, Hu and Zhou (2022) formulate the problem as a stochastic DP and propose heuristic policies to compute the optimal matching decisions. For a related problem, the truckload carriers assignment problem, Simao et al. (2009) also formulate a dynamic DP but with post-decision states so that they are able to solve the problem using ADP. In each iteration, a demand path is sampled, and the value function is approximated in a linear form and updated using the dual variables from the LP solution to the resulting optimization problem.

5.1   Optimization Formulation

To facilitate the formulation and statement of the optimization problem, we summarize the quantities in the paper and their notation in Table 5.1.

Our optimization horizon is 24 h. A trip order can be summarized as $o := \{\tilde{l}_o, l_o, l_d, t_r, t_m, \tilde{t}_o, t_o, t_d, p\}$. The tilde for $\tilde{l}_o$ and $\tilde{t}_o$ indicates that they have additional dependencies on

Z. (Tony). Qin et al., *Reinforcement Learning in the Ridesharing Marketplace*, Synthesis Lectures on Learning, Networks, and Algorithms, https://doi.org/10.1007/978-3-031-59640-7_5

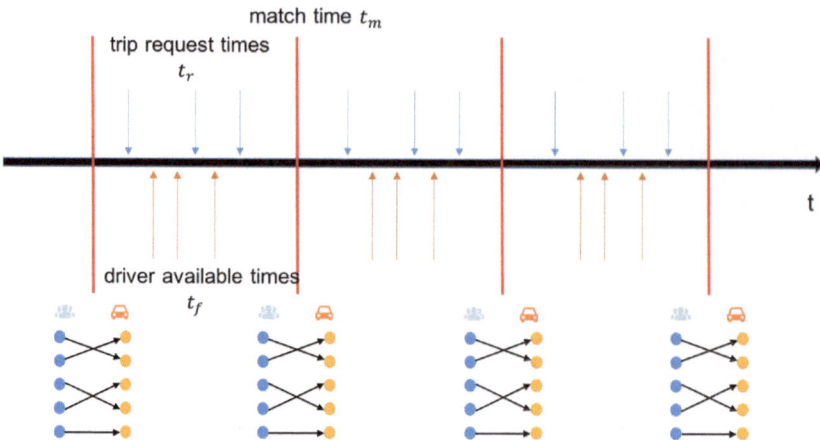

Fig. 5.1 The order matching process with batching from the system perspective (Qin et al. 2020a). The assignments are for illustration only

the assigned driver. We use i to index orders as in $o^{(i)}$. A driver x is represented by $x := \{t_f, l_x, t_x\}$, where t_f is the time when the driver last became available, and (l_x, t_x) is the current spatiotemporal state of the driver. If the driver is in service, $t_f = \infty$. The duration of the trip $o^{(i)}$ is $\tau_o^{(i)} := t_d^{(i)} - t_o^{(i)}$. The time that the driver spends en route to pick up the passenger is $\tau_e^{(i)} := t_o^{(i)} - \tilde{t}_o^{(i)}$. The total time for fulfilling the order is thus $\tau_o^{(i)} + \tau_e^{(i)} = t_d^{(i)} - \tilde{t}_o^{(i)}$. An order o is eligible to be assigned to a driver x if $t_m(o) \geq t_f$, where $t_m(o)$ is the t_m component of o (and this notational convention applies to the other quantities as well). For simplicity, we shorten the notation $t_m(o^{(i)})$ to $t_m^{(i)}$. We use j to index drivers as in $x^{(j)}$. We denote the set of free drivers by $X(t) := \{x^{(j)} \mid t_f^{(j)} \leq t\}$ and the global state at t by $\tilde{S}(t) := (X(t), O_t)$, where O_t is the set of open orders at t. An order-dispatching policy π is a function that maps an order o to a free driver $x \in X(t_m(o))$ given $\tilde{S}(t_m(o))$, that is,

$$\pi(o; \tilde{S}(t_m(o))) : o \to x \in X(t_m(o)). \tag{5.1}$$

It is understood that if multiple orders have the same dispatch assignment time t_m, then the dispatching policy π ensures that two orders are not matched to the same free driver. If $X(t_r(o)) = \emptyset$, then the order is not matched, and its dispatch request time is advanced to the new decision time until $X(t_r(o))$ is nonempty. We set the price $p = 0$ if the order is cancelled before being fulfilled. A cancellation after an order has been assigned is usually because of a long pickup distance or a long estimated pickup wait time $\hat{t}_o^{(i)} - \tilde{t}_o^{(i)}$, defined by $d(\tilde{l}_o^{(i)}, l_o^{(i)})$, where the function d returns the travel distance between two locations. Hence, p is a function of π, and we can make it explicit by writing $p(\pi)$.

Table 5.1 We List the Core Mathematical Notation We Use in this Paper. Notations not listed in this table are explicitly defined and explained in this case study when they are presented

Symbol	Meaning
o	Order object
l_o	Trip origin in coordinates
l_d	Trip destination in coordinates
p	Actual trip price
$\hat{p}$	Trip price quote
t_r	Order submission time
t_m	Trip assignment (to the driver) time
$\tilde{t}_o$	Driver acceptance (of the assignment of order o) time
$\tilde{l}_o$	Driver's location at assignment (of order o) acceptance
$\hat{t}_o$	Estimated pickup time
$\hat{t}_d$	Estimated drop-off time
t_o	Actual pickup time
t_d	Actual drop-off time
O_{disp}	Set of open orders within a batch window
x	Driver object
t_f	Time when driver last became available
(l_x, t_x)	Current spatiotemporal state of driver
X_{disp}	Set of available drivers within a batch window

5.2   Metrics

The RL literature for rideshare matching (see Table B.2) typically aims to optimize the total driver income and the service quality over an extended period of time. Service quality can be quantified by *response rate* and *fulfillment rate*. Response rate is the ratio of the matched requests to all trip requests. Since the probability of pre-match cancellation is primarily a function of response time (pre-match waiting time), the total response time is an alternative metric to response rate. Fulfillment rate is the ratio of completed requests to all requests and is no higher than the response rate. The gap is due to post-match cancellation, usually because of the waiting for pick-up. Hence, the average pick-up distance is also a relevant quantity to observe. Figure 5.2 shows the detailed flow of matching a single trip request together with the quantities discussed above.

Let the set of orders created by passengers throughout a day be $\{o^{(i)}\}_{i=1}^{N}$. We have both driver-centric and passenger-centric objectives introduced above. The driver-centric objective is to maximize the total income of the drivers on the platform. For simplicity, the

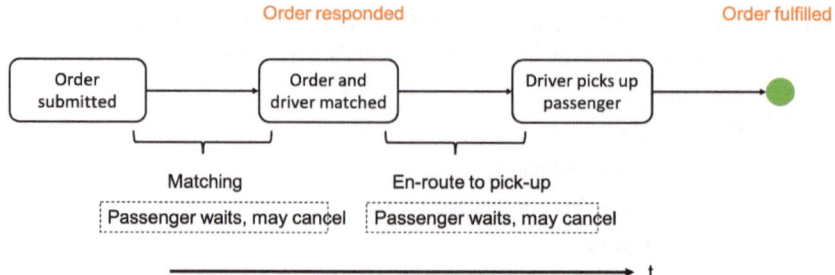

Fig. 5.2 The order matching process from a single request's perspective

per-order income of a driver is $r := p\theta$, where θ is a constant independent from the order. Hence, the optimization problem is

$$\max_{\pi} \ J(\pi) := \sum_{i=1}^{N} p^{(i)}(\pi). \tag{5.2}$$

The passenger-centric objective is to minimize the average pickup distance of all the assigned orders, $\frac{1}{N^+} \sum_{i=1}^{N} (d(\tilde{l}_o^{(i)}, l_o^{(i)}))_{\mathbf{1}_{(\tilde{l}_o^{(i)} \neq \emptyset)}}$, where N^+ is the number of assigned orders. This manages the passenger experience in terms of the waiting time for pickup. Response rate is defined as the percentage of all submitted orders that are assigned to a driver, $\frac{N^+}{N}$. Fulfillment rate is the percentage of all submitted orders that are eventually completed, $\frac{\sum_{i=1}^{N} \mathbf{1}(p^{(i)}(\pi)>0)}{N}$.

5.3   Single-Agent Formulation

In terms of the MDP formulation, driver agent is a convenient modeling choice for its straightforward definition of state, action, and reward, in contrast to system-level modeling where the action space is exponential. In this case, the rideshare platform is naturally a multi-agent system with a global objective. A common approach is to crowdsource all drivers' experience trajectories to train a single agent and apply it to all the drivers to generate their matching policies (Xu et al. 2018; Wang et al. 2018; Tang et al. 2019). Since the system reward is the sum of the drivers' rewards, the system value function does decompose into the individual drivers' value functions computed by each driver's own trajectories. The approximation here is using a single value function learned from all drivers' data. See Qin et al. (2020a) for detailed discussions. Specifically, Xu et al. (2018) learn a tabular driver value function using TD(0), and Wang et al. (2018), Tang et al. (2019), Holler et al. (2019) apply DQN-type of training to learn a value network. In particular, Tang et al. (2019) design a spatiotemporal state-value network using hierarchical coarse coding and cerebellar embedding memories for better state representation and training stability.

This type of single-agent approach avoids dealing explicitly with the multi-agent aspect of the problem and the interaction among the agents during training. Besides simplicity, this strategy has the additional advantage of being able to easily handle a dynamic set of agents (and hence, a changing action space) (Ke et al. 2020b). On the other hand, order matching requires strong system-level coordination in that a feasible solution has to satisfy the one-to-one constraints. To address this issue, Xu et al. (2018), Tang et al. (2019) use the learned state values to populate the edge weights of a bipartite assignment problem to generate a collective-greedy policy (Qin et al. 2020a) with respect to the state values. Holler et al. (2019) assume a setting where drivers are matched or repositioned sequentially so that the policy output always satisfies the matching constraints.

Earlier deployed works, e.g., Xu et al. (2018), Tang et al. (2019), Qin et al. (2020a), adopt offline learning to learn the driver values, which are used in solving the assignment problem online. More recently, several works (Tang et al. 2021; Eshkevari et al. 2022; Han et al. 2022; Azagirre et al. 2023) have demonstrated the feasibility and superiority of online on-policy value updates for the dispatching problem in real-world settings. We will elaborate on this exciting new development in Sect. 5.9.

5.4   Multi-agent and System-Level Formulation

Because of the combinatorial nature of the action space of a system agent, system-level modeling for the rideshare matching problem often adopts a sequential matching policy. It can be to explicitly assume a sequential matching mechanism within which there is no batching window, and the requests are matched on a first-come-first-serve basis, or to implicitly serialize an otherwise batched decision. Under this setting, the decision at any time involves only one request. Holler et al. (2019) develop an action-value network that leverages global SD information, which is embedded into a global context by attention. The approach adopted by Feng et al. (2020) is an example of serialization, where a system-level batch assignment is solved by determining the assignment for each request sequentially.

Leveraging MARL, Li et al. (2019), Jin et al. (2019), Zhou et al. (2019) directly optimize the multi-agent system. One significant challenge is scalability since any realistic ridesharing setting can easily involve thousands of agents, precluding the possibility of dealing with an exact joint action space. Li et al. (2019) apply mean-field MARL to make the interaction among agents tractable, by taking the 'average' action of the neighboring agents to approximate the joint actions. Zhou et al. (2019) argue that no explicit communication among agents is required for order matching due to the asynchronous nature of the transitions and propose independent Q-learning with centralized KL divergence (of the supply and demand distributions) regularization. Both Li et al. (2019), Zhou et al. (2019) follow the centralized training decentralized execution paradigm. Jin et al. (2019) take a different approach treating each spatial grid cell as a worker agent and a region of a set of grid cells as a manager agent, and they adopt hierarchical RL to jointly optimize order matching and vehicle repositioning.

5.5   Multi-objective

In practical settings, the online matching policy often has to balance among multiple objectives (Lyu et al. 2019), e.g., financial metrics and customer experience metrics. The rationale is that persistent negative customer experience will eventually impact long-term financial metrics as users churn the service or switch to competitors. There are two potential ways that one can leverage RL to approach this problem. The explicit approach is to directly learn a policy that dynamically adjusts the weights to combine the multiple objectives into a single reward function. With the abundance of historical experience data, inverse RL can be used to learn the relative importance of multiple objectives under a given unknown policy (Zhou et al. 2021a). The implicit approach is to capture the necessary state signals that characterize the impact of the metrics not explicitly in the reward function, so that the learned value function correctly reflect the long-term effect of the multiple metrics. As discussed in Sect. 11.5, the long feedback loop is a potential challenge here.

5.6   Other Decisions

Besides the driver-passenger pairing decisions, there are other important levers that can be optimized within the matching module, namely the matching window and the matching radius (Yang et al. 2020a). The matching window determines when to match a request (or a batch of requests). A larger window increases pre-match waiting time but may decrease pick-up time for matched requests because of more available drivers. There have been several RL works on the matching window optimization, which can be done from the perspective of a request itself (Ke et al. 2020b) or the system (Wang et al. 2019; Qin et al. 2021a). In Ke et al. (2020b), each trip request is an agent. An agent network is trained centrally using pooled experience from all agents to decide whether or not to delay the matching of a request to the next review window, and all the agents share the same policy. To encourage cooperation among the agents, a specially shaped reward function is used to account for both local and global reward feedback. They also modify the RL training framework to address the delayed reward issue by sampling complete trajectories at the end of training epochs to update the network parameters. Wang et al. (2019) take a system's view and propose a Restricted Q-learning algorithm to determine the length of the current review window (or batch size). They show theoretical analysis results on the performance guarantee in terms of competitive ratio for dynamic bipartite graph matching with adaptive windows. Qin et al. (2021a) take a similar modeling perspective but use the AC method with experience replay (ACER) (Wang et al. 2016) that combines on-policy updates (through a queuing-based simulator) with off-policy updates. The matching radius defines how far an idle driver can be from the origin of a given request to be considered in the matching. It can be defined in travel distance or time. A larger matching radius may increase the average pick-up distance but requests are more likely to be matched within a batch window, whereas a smaller radius renders less effective driver availability but it may decrease the average pick-up distance.

Both the matching window and radius are trade-offs between pre-match and post-match waiting times (and hence, cancellation). So far, few effort through RL has been devoted to matching radius optimization. The joint optimization of the matching window and radius is certainly another interesting line of research.

5.7   Matching in Related Domains

Because of its generalizability, matching for ridesharing is closed related to a number of online matching problems in other domains, the RL methods to which are also relevant and can inform the research in rideshare matching. Some examples are training a truck agent using DQN with pooled experience to dispatch trucks for mining tasks (Zhang et al. 2020b), learning a decentralized value function using PPO with a shaped reward function for cooperation (in similar spirit as Ke et al. (2020b)) to dispatch couriers for pick-up services (Chen et al. 2019b), and designing a self-attention, pointer network-based policy network for a system agent to assign participants to tasks in mobile crowdsourcing (Shen et al. 2020).

5.8   Case Study: Batch Learning

We focus on the line of works for learning the driver-level value function for the matching problem. We generally follow chronological order—first describing the batch learning methods (Xu et al. 2018; Qin et al. 2020a; Tang et al. 2019), and then discussing the more recent online RL algorithms (Tang et al. 2021; Eshkevari et al. 2022) to solve this problem in Sect. 5.9.

5.8.1   Semi-MDP Formulation

In this model, each driver is an independent agent. The state s of the driver consists of location and time (l, t), both of which can be discretized: the driver's location is represented in a hexagonal grid system (e.g., Brodsky 2018), and time is represented by buckets, typically of a few minutes. A state is a *terminal* state, if $t = T$ where T is the end time of a day, or an episode. We denote the (hexagonal grid index, time bucket index) pair of location l and time t by $g(l, t)$. The hexagonal grid system is commonly used in mapping systems because it has a desirable property that the Euclidean distance between the center points of every pair of neighboring grid cells is the same, and hexagonal grids have the optimal perimeter/area ratio, which leads to a good approximation of circles (Hales 2001). The action a of the driver (or rather the action that the system imposes on the driver) is to fulfill a particular order (from the open orders within a matching window), or to idle. The reward r of an action executed on a given state is simply the price of the order p, which can be zero if the driver is idle. The state transition dynamics are that after the driver at $s = g(\tilde{l}_o, \tilde{t}_o)$ completes an order o, the driver's state changes to $s' = g(l_d, t_d)$, and the driver receives a reward of $r = p$. Hence, both the transition and reward are deterministic, given s and a. A sample trajectory of the driver in an episode is shown in Fig. 5.3. The stochasticity of this MDP lies in the future demand, which defines the feasible action set at each state. Hence, strictly speaking, our MDP is one with stochastic action sets (SAS-MDP) (Boutilier et al. 2018). Regular learning algorithms like Q-learning, DQN, and policy evaluation methods still work the same in this case using batch data, as long as updates are made over realized available actions. The objective of this MDP is to maximize the cumulative reward of the agent (driver) x within an episode, $J_x = \sum_{k=1}^{K} r_{t_k}$, where t_k is the time of the k-th action, and t_K is the time of the last action before T. For notational simplicity, we subsequently may just use r_k for r_{t_k}. Since all idle actions yield zero reward, we have

$$J_x(\pi) = \sum_{k=1}^{K} r_{t_k} = \sum_{i=1}^{N} p^{(i)}(\pi)\bigg|_{\pi(o^{(i)})=x}. \qquad (5.3)$$

From Eq. (5.3), it follows that

$$J(\pi) = \sum_{x} J_x(\pi). \qquad (5.4)$$

This model contains temporally extended courses of actions, so it is in fact a semi-MDP as Tang et al. (2019) describe, and the actions are options (Sutton et al. 1999), which we also denote by o (and is consistent with its meaning). Most relevant theories of MDP can be

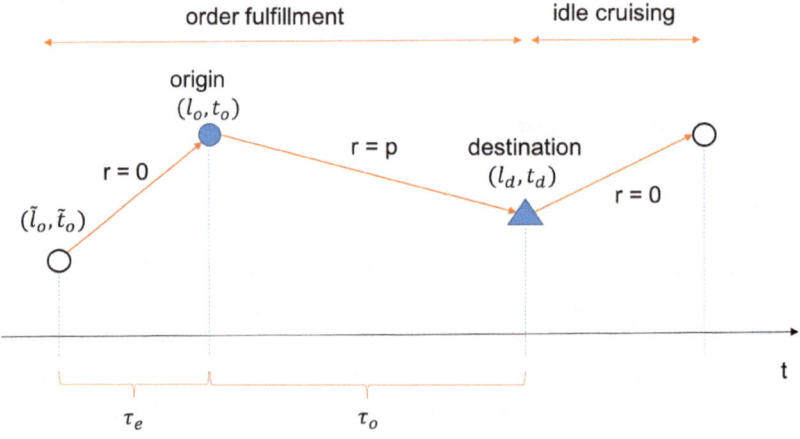

Fig. 5.3 The Graphic Shows a Sample Driver Trajectory
Notes The first two legs correspond to a transition for order fulfillment (i.e., pickup plus actual trip). The last leg is a transition for idle cruising. *Source* Qin et al. (2020a)

carried over with only minor modification (Sutton et al. 1999). The reward r accumulated over an option whose trip portion spans over τ_o units of time needs to be properly discounted:

$$\hat{r} = \gamma^{\tau_e}\left(\frac{r}{\tau_o} + \gamma\frac{r}{\tau_o} + \cdots + \gamma^{\tau_o-1}\frac{r}{\tau_o}\right) = \frac{r(\gamma^{\tau_o}-1)\gamma^{\tau_e}}{\tau_o(\gamma-1)}, \tag{5.5}$$

where $\gamma \in [0, 1)$, $\tau_o \geq 1$ and assuming that the reward is accumulated uniformly over time. Here, we have also taken into account the time that the driver spent en route to pick up the passenger, τ_e. The policy of the agent π_d is a function that maps the driver's state to an option, that is, $\pi_d(s) : S \rightarrow O$, where S and O are the state and option spaces, respectively. π can be distilled into π_d through

$$\pi_d(s(x)) = o^{(i)} \text{ iff } \pi(o^{(i)}) = x. \tag{5.6}$$

From (5.1), it is understood that the system policy π is also conditional on the global state. The episode trajectory data of driver x is $P^{(x)} = \{s_{t_0}, o_0, r_0, \ldots, s_{t_k}, o_k, r_k, s_{t_{k+1}}, \ldots, s_{t_K}\}$. We use the notation o_k for the k-th option at time t_k to differentiate it from the indexing for orders $o^{(i)}$. The option o_k may be to idle in addition to taking a trip.

Similarly for an MDP, the state-value function $V^{\pi_d}(s)$ of the semi-MDP is defined as the long-term discounted cumulative reward received throughout the options given s, following π_d:

$$V^{\pi_d}(s) := E\left[\sum_{i=1}^{K-k} \gamma^{(t_{k+i}-t_k-\tau_{o_k}-\tau_{e_k})}\hat{R}_{k+i} \,\middle|\, s_{t_k} = s\right], \tag{5.7}$$

where $\hat{R}$ is the random variable whose realized value is $\hat{r}$. The time component for the discount factor in front of $\hat{R}_{k+i}$ is because the reward $\hat{R}_{k+i}$ starts to be collected after the k-th transition is completed. The Bellman equation for π_d is

$$V^{\pi_d}(s) = E_{O_{disp}(s)\sim\mathcal{O}}\left[\hat{R}(s, \pi_d(s; O_{disp}(s))) + \gamma^{(\tau_e+\tau_o)}V^{\pi_d}(s')\right], \tag{5.8}$$

where $O_{disp}(s)$ is the set of open trip orders for dispatching (i.e., the action set) at s, and $\mathcal{O}$ is the corresponding demand distribution. We made the dependency of π_d on O_{disp} explicit by writing $\pi_d(s; O_{disp}(s))$. Following (Boutilier et al. 2018), we can define the corresponding embedded semi-MDP by augmenting the state s with the realized action set $O_{disp}(s)$, $\tilde{s} := (s, O_{disp}(s)) \in \tilde{S}$ and denoting the corresponding policy $\tilde{\pi}_d$. Then we can recover the standard Bellman equation for $\tilde{\pi}_d$,

$$V^{\tilde{\pi}_d}(\tilde{s}) = E_{\tilde{s}'\sim\tilde{S}}\left[\hat{R}(\tilde{s}, \tilde{\pi}_d(\tilde{s})) + \gamma^{(\tau_e+\tau_o)}V^{\tilde{\pi}_d}(\tilde{s}')\right], \tag{5.9}$$

where $\tilde{S}$ is the conditional distribution of the state $\tilde{s}'$ given the action $\tilde{\pi}_d(\tilde{s})$. In practice, we use TD-learning to learn $V^{\pi_d}(s)$ from a collection of realized trajectories.

Similarly, the state-option value function $Q^{\pi_d}(s, o)$ is defined as the long-term discounted cumulative reward received throughout the options by following π_d given current state s and executing option o on s:

$$Q^{\pi_d}(s, o) := E\left[\sum_{i=1}^{K-k} \gamma^{(t_{k+i} - t_k - \tau_{o_k} - \tau_{e_k})} \hat{R}_{k+i} \,\middle|\, s_{t_k} = s, o_k = o\right]. \tag{5.10}$$

5.8.2 Value Network

At the policy evaluation stage, V^{π_d} is learned through tabular temporal-difference (TD) learning (Sutton 1988) (TD(0) to be exact) using $\{P^{(x)}\}_{x=x^{(0)}}^{x^{(J)}}$, the trip and idle movement data for all drivers (from driver $x^{(0)}$ to $x^{(J)}$) collected for the training period, for example, a month. Algorithm 5.1 lays out the key steps of TD(0) applied to the semi-MDP in the previous section. The update term $r + \gamma V^{\pi_d}(s') - V^{\pi_d}(s)$ is the *TD-error* for the transition experience (s, o, r, s'), and α is the step size.

Algorithm 5.1 TD(0) for Driver Semi-MDP

Require: $\{P^{(x)}\}_{x=x^{(0)}}^{x^{(J)}}$ for the training period collected by π, $\alpha \in (0, 1]$
Ensure: $V^{\pi_d}(s) = 0$, $\forall s$ whose time-bucket contains T.
  Initialization: $V^{\pi_d}(s) = 0$, $\forall s \in S$
  for each episode P **do**
    for each transition (s, o, r, s') **do**
      $V^{\pi_d}(s) \leftarrow V^{\pi_d}(s) + \alpha\left[\hat{r} + \gamma^{(\tau_o + \tau_e)} V^{\pi_d}(s') - V^{\pi_d}(s)\right]$
    end for
  end for

The improved system dispatching policy π' with respect to V^{π_d} is generated by the operator $\Pi(w)$ during the matching stage through combinatorial optimization (linear assignment problem solved by the Hungarian method (Kuhn 1955)). The edge weights are computed as the sample (predicted) advantage (Baird III 1993) of each possible match between $o^{(i)}$ and $x^{(j)}$, using V^{π_d}:

$$w_{o^{(i)}, x^{(j)}}(V^{\pi_d}) := \hat{p}^{(i)} + \gamma^{(\hat{\tau}_o^{(i)} + \hat{\tau}_e^{(i)})} V^{\pi_d}(g(l_d^{(i)}, \hat{t}_d^{(i)})) - V^{\pi_d}(s(x^{(j)})), \tag{5.11}$$

with the understanding that $\hat{p}^{(i)}$ is discounted as in Eq. (5.5). The advantage, Eq. (5.11), can be viewed as the relative change in the long-term value with respect to the current spatiotemporal point of the driver $x^{(j)}$, should order $o^{(i)}$ be assigned to $x^{(j)}$. The generated policy $\pi' = \Pi(w(V^{\pi_d}))$ is collective-greedy with respect to V^{π_d} (through the advantage). The sample advantage admits the same form as the TD-error. We observe that the edge weight, Eq. (5.11), penalizes long pickup distance in that both the immediate reward $\hat{p}^{(i)}$ (see Eq. (5.5)) and the discount factor for the value term of the destination would decrease with increasing estimated en route time $\hat{\tau}_e^{(i)}$, thus lowering the advantage. The new policy is guided by the independent long-term option advantages of the drivers to approximately maximize their total income while discouraging long pickup wait times for the passengers. We formalize our RL framework for dispatching in Algorithm 5.2.

Algorithm 5.2 Generalized Policy Iteration for Order Dispatching

---

Require: Dispatching policy π (and corresponding π_d). Storage buffer B with episode trajectory data $\{P^{(x)}\}$ collected by π.
  for $t = 1, 2, \ldots$ **do**
    Learn the value function V^{π_d} from the data in B using a policy evaluation method, e.g., CVNet, TD(0).
    Compute $w(V^{\pi_d})$ by (5.11) and generate $\pi' = \Pi(w)$.
    Match orders and drivers with batch windows using π'.
    Collect new trip and driver trajectory data. Fill B with new data.
    $\pi \leftarrow \pi'$
  end for

Tabular TD-learning has enabled us to improve the system dispatching policy with respect to the average long-term values of individual drivers, but it also has several limitations. (1) Tabular methods suffer from the curse-of-dimensionality. As the number of features to represent the agent state increases, the table size for the value function quickly becomes intractable. (2) The tabular TD method is susceptible to training data sparsity because it is unable to generalize in a principled way to spatiotemporal states that have not been visited by any driver in the past. (3) Tabular learning methods do not support the mechanism for knowledge sharing among models of different cities. Meanwhile, using merely spatiotemporal information is not sufficient to capture the complex nature of the driver state. The state space needs to be augmented to enable policies to be more responsive to real-time demand and supply conditions and to better accommodate driver heterogeneity. The dispatching system has to support potentially hundreds of cities with very different data availability. The training method thus needs to be able to leverage knowledge sharing among models of different cities to reduce training time and improve learning quality. With all these considerations in mind, we developed a deep neural network-based policy evaluation algorithm, the cerebellar value-network (CVNet) (Tang et al. 2019), for learning the driver-centric state value function for our generalized policy iteration framework. The development of a CVNet for multidriver order dispatching requires several innovative features in network design and training.

The state values of similar geographical locations at a given time tend to be similar, but due to different densities of the transportation network and natural geographical features, the grouping size of the spatial points can vary. Hence, neither a single-resolution grid system nor the raw GPS coordinates are optimal for state representation and learning. CVNet quantizes the geographical space through hierarchical coarse-coding using a multiresolution hexagonal grid system (Sahr 2011; Brodsky 2018; Uher et al. 2019). State representation is then constructed by combining a Cerebellar Model Arithmetic Computer (CMAC) (Albus 1971) with an embedding matrix.

An input spatial point l to the CMAC activates a set of grids of multiple resolutions by the quantization functions $\{q_k(l)\}_k$, which generate a sparse activation vector $c(l) \in \mathbb{R}^A$ (A is the size of *conceptual memory*) that maps l to appropriate rows of the embedding matrix $M \in \mathbb{R}^{A \times m}$ by $c(l)^T M$. The embeddings are updated as part of the neural network in the learning (training) stage to learn the best feature representation of each grid. The embedding layer in conjunction with other state features (e.g., supply-demand context) is connected to a multilayer perceptron (MLP) to output the final state value. CVNet is trained within a DQN-like framework with minibatch stochastic gradient decent (Fig. 5.4).

Sensitivity of the value function to input perturbation would propagate to the policy derived from the value function. To improve the robustness of CVNet to input perturbation (to which a tabular value function is susceptible when training data are sparse for certain parts of the spatiotemporal space, creating 'spikes' in the value table), we regularize an upper bound of the Lipschitz constant of V^{π_d}, $\mathcal{L}(V^{\pi_d})$, that is, to bound the output with respect to the norm of all input states. Since $V^{\pi_d}(s) = (v_L \circ v_{L-1} \circ \cdots v_1)(s)$, where $\{v_h\}_{h=1}^L$ are a series of constituent functions for the L layers of the neural network, $\mathcal{L}(V^{\pi_d}) \leq \Pi_h \mathcal{L}(v_h)$, and we have derived the analytical forms of the Lipschitz constants for the Cerebellar Embedding layer and the MLP layers in Tang et al. (2019).

CVNet was deployed in production for AB tests on a major ridesharing platform and showed significant improvement of $0.5-2\%$ against the baseline in key marketplace metrics, such as total driver income, order response rate, and fulfillment rate (Qin et al. 2020a).

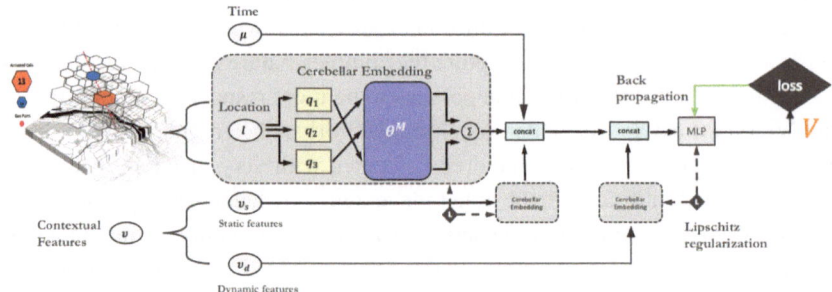

Fig. 5.4 The diagram shows the CVNet architecture as (Tang et al. 2019) discuss. The left side and center of the diagram illustrate the hierarchical coarse-coding using a multiresolution hexagonal grid system and cerebellar embedding. q_1, q_2, and q_3 represent the quantization functions, and θ^M denotes the parameters of the embedding matrix defined in Sect. 5.8.2

5.8.3 Transfer Learning

One of the key advantages of employing deep reinforcement learning is to allow leveraging knowledge learned from one city in training to improve training efficiency and quality for other cities. For transfer learning of the models among different cities, we further refined our neural network into a dual-pathway architecture (Wang et al. 2018; Tang et al. 2019), where there are two trains of network layers corresponding to location features (l) and transferrable features such as time (μ), spatiotemporal displacement and local supply-demand contextual features (v). The two pathways are connected thru lateral connections. Figure 5.5 illustrates this network architecture, which is called the Correlated Feature Progressive Transfer (CFPT) architecture. Once the network for the source city is trained, the transferrable blocks of the network are transplanted to the right positions in the target network, the nontransferrable blocks of which continue to be updated by the new data from the target city. In one set of experiments (Tang et al. 2019) we demonstrate the effect of transfer learning by running simulations across multiple days and cities using real-world data collected from a major ride-hailing platform, in which we run simulations under different pickup distance penalties and obtain a trade-off between Total Driver Income (TDI) and the pickup distance. It is observed in the experiments that using transfer learning attains a greater improvement on TDI while allowing wider range of pickup distance options, compared with the Baseline with no transfer. In particular, the method CFPT proposed in Tang et al. (2019) is shown to be especially effective when applied to cities where training data is not as abundant, e.g., cities with smaller population, while achieving competitive and consistent performance in other cases compared with Finetuing (Hinton and Salakhutdinov 2006) and Progressive (Rusu et al. 2016).

5.8.4 Why Offline Learning Works

Recent results from Brandfonbrener et al. (2021) provides an excellent explanation and justification for the approach introduced in this case study. They show that simply doing one step of constrained/regularized policy improvement using an on-policy evaluation of the behavior policy performs surprisingly well and outperforms existing iterative offline RL algorithms based on off-policy evaluation. Although proposed several years earlier, the batch learning methods in this case study are instances of the on-policy evaluation, one-step improvement method proposed in Brandfonbrener et al. (2021) because the value network obtained from offline batch learning is used for dispatch execution in production for a period of time (e.g., a week) before it is re-learned from the updated data set. So, each learning-deployment iteration essentially runs the method in Brandfonbrener et al. (2021).

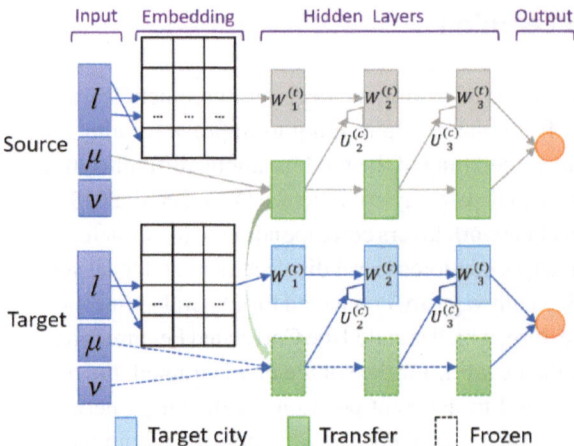

Fig. 5.5 The Diagram Shows the CFPT Architecture for Transfer Learning (Tang et al. 2019) *Notes l*, μ, and u are the location, time, and contextual features, respectively. The blue blocks in the upper pathway of the target network are the network layers specific to the target city. The green blocks in the lower pathway of both networks are transferrable blocks between the source and target cities. $W^{(t)}$ and $U^{(c)}$ are the network weights of the nontransferrable blocks and the lateral connections, respectively. The parts of the target city network with dotted lines (frozen) are not updated during training

5.9   Case Study: Online On-Policy Learning in Ridesharing

The RL track in the KDD Cup 2020 competition (Qin et al. 2020b) marked a significant milestone in the development of RL methods in ridesharing. The competition revealed the potential of online on-policy updates to outperform the offline-trained value functions, despite previous concerns about convergence and stability. This development came after a period where the RL methods used in ridesharing deployment primarily relied on offline batch learning. The online on-policy learning approach demonstrated its effectiveness in a realistic simulation environment during the competition. This chapter seeks to delve into the online on-policy learning methods that were implemented following the competition.

5.9.1   Joint Online Optimization Through Value-Based Methods

A comprehensive learning framework, referred to as **V1D3**, leverages **V**alue-based dynamic learning for order **D**ispatching and **D**river repositioning (Tang et al. 2021). This method is characterized by its capacity to rapidly adapt to real-time dynamics and maintain efficacious operational levels in large-scale, real-world scenarios encompassing tens of thousands of agents (vehicles).

V1D3's superiority was ascertained through methodical experiments using a ride-hailing simulation environment based on real-world data, as part of the KDD Cup 2020 RL competition. The results demonstrated that V1D3 significantly outpaced state-of-the-art methods, including the top-performing models of the order dispatching and repositioning tracks in the competition.

Developing a unified framework for both tasks that can scale to large, real-world scenarios presents significant challenges. To achieve the best performance, the framework must take proper account of the interactivity between the dispatching and repositioning tasks. For example, the dispatching task needs to consider the repositioning results which affect supply distributions. Conversely, the repositioning task must factor in the dispatching probability, which changes with the dispatching policy. However, models trained offline that rely on contextual dependencies and real-time inputs have practical limitations, especially when multiple such models interact in real time.

Moreover, the large-scale nature of both tasks necessitates efficient coordination among agents to address dispatching constraints and prevent undesirable competition amongst vehicles. For instance, a straightforward application of multi-agent Deep Reinforcement Learning (Tampuu et al. 2017; Lowe et al. 2017a), would only permit coordination among a restricted set of agents due to the overwhelming computational costs.

Further, the framework needs to handle daily variations in supply and demand in ridesharing marketplaces. It should also be capable of adapting to unexpected events, such as spikes in demand due to a significant event like a concert, which cause significant short-term fluctuations in a particular area.

5.9.1.1 Online Updates Within a Population Context

The value function updates should ideally capture the ever-changing dynamics of supply-demand conditions in real-time. Consider the scenario where a set of available drivers denoted by $\mathcal{D}$, are updated based on the actions they execute in each dispatch period. This means accounting for every driver's unique transition. For instance, for every driver i who was successfully assigned an order from the set $\mathcal{D}_D$, their current state s^i_{driver} and the final state s^i_{order} of the assigned order can be represented using the one-step Bellman update as follows:

$$V(s^i_{driver}) \leftarrow r^i_{order} + \gamma^{\Delta t_{order}} V(s^i_{order}) \qquad (5.12)$$

Here, r^i_{order} represents the corresponding order trip fee, and Δt_{order} denotes the estimated order trip duration.

For every idle driver i in $\mathcal{D}_I$, not assigned an order during a particular dispatch cycle, their current state s^i_{driver} and their subsequent state after idling s^i_{idle} can be updated with Bellman update:

$$V(s^i_{driver}) \leftarrow 0 + \gamma^{\Delta t_{idle}} V(s^i_{idle}) \qquad (5.13)$$

The transition here incurs no reward and lasts for Δt_{idle} duration.

Relying on strategies used in practical Q-learning (e.g., Van Hasselt et al. 2016; Mnih et al. 2015), these Bellman updates can be converted into a bootstrapping-based objective for training a V-network, V_θ, through gradient descent. This objective is commonly referred to as mean-squared temporal difference (TD) error. Specifically, let δ_θ^i denote the TD error for the ith driver, we get:

$$\delta_\theta^i = \begin{cases} r_{order}^i + \gamma^{\Delta t_{order}} V_\theta(s_{order}^i) - V_\theta(s_{driver}^i) & \forall i \in \mathcal{D}_D; \\ \gamma^{\Delta t_{idle}} V_\theta(s_{idle}^i) - V_\theta(s_{driver}^i) & \forall i \in \mathcal{D}_I. \end{cases} \tag{5.14}$$

When applied to all drivers in $\mathcal{D}$, this gives us the *population-based mean-squared TD error*:

$$\min_\theta \ L(\mathcal{D}; \theta) := \sum_{i \in \mathcal{D}_D} (V_\theta(s_{driver}^i) - r_{order}^i - \gamma^{\Delta t_{order}} \bar{V}_\theta(s_{order}^i))^2$$

$$+ \sum_{i \in \mathcal{D}_I} (V_\theta(s_{driver}^i) - \gamma^{\Delta t_{idle}} \bar{V}_\theta(s_{idle}^i))^2 = \sum_{i \in \mathcal{D}} (\delta_\theta^i)^2 \tag{5.15}$$

Here, a target network $\bar{V}_\theta$ is used to stabilize the training following common practices (Van Hasselt et al. 2016). The target network acts as a delayed copy of the V-network V_θ. After each round of dispatch, V_θ is updated by taking a gradient descent step aimed at minimizing $L(\theta)$. This operation can be represented as $\theta \leftarrow \theta - \alpha \nabla L(\mathcal{D}; \theta)$ where $\alpha > 0$ is a step-size parameter that modulates the learning rate.

5.9.1.2 Unified Learning Framework

The unified learning framework makes use of the updated value function V_θ obtained through online on-policy updates. This function addresses coordination challenges of both multi-task (dispatching and repositioning) and multi-agent (driver) situations effectively.

In matching, the matching policy is generated following the method described in Sect. 5.8.2. Here, the utility of each dispatch graph edge is determined by the corresponding potential match's TD-error or advantage. From a multi-agent environment perspective with thousands of drivers, solving the dispatch graph can be seen as a policy improvement step (Sutton et al. 1998).

Repositioning is considered for the set of repositioning drivers, denoted $\mathcal{I}$. This set contains all drivers with idle time exceeding a threshold of C minutes (usually between five and ten minutes). For each driver i in $\mathcal{I}$, repositioning the driver to a location selected from a set of candidate destinations, $\mathcal{O}_d(s^i)$, given the driver's current state s^i maximizes their expected long-term return. Specifically, we sample the reposition destination with a probability proportional to the discounted state value function as:

$$p(s_k^i) \sim \frac{e^{\gamma^{\Delta t_{ik}} V_\theta(s_k^i)}}{\sum_{j \in \mathcal{O}_d(s^i)} e^{\gamma^{\Delta t_{ij}} V_\theta(s_j^i)}}, \quad \forall k \in \mathcal{O}_d(s^i) \tag{5.16}$$

Here, $0 < \gamma \leq 1$ is the discount factor, and Δt_{ik} denotes the estimated travel time to the destination k. It is pertinent to note that the driver's current location is always included in the candidate set $i \in \mathcal{O}_d(s^i)$. Therefore, the travel time will be $\Delta t_{ii} = 0$, and the state value will not be discounted when computing the sampling probability. Hence, the cost of repositioning to a state different from the current one ensures a closer destination with a smaller reposition cost is preferred, given the same state value. We will further discuss in Sect. 6.4.2 that this stochastic policy, based on sampling from the distribution (5.16), achieves significantly better group-level performance than the greedy policy for a large fleet.

The unified learning framework comprises of a shared current state value network V_θ employed to reposition idle drivers and dispatch outstanding orders at each time step. Constraints are taken care of by solving an assignment problem that maximizes the total advantage of the driver population with a view to matching. The framework triggers repositioning every $C \approx 150$ time steps but only for those drivers who have been idle for over C time steps.

This framework can effectively scale to a large number of reposition drivers with minimal performance losses despite its complexity. This capability is made possible by two key design elements. To begin with, the value network is frequently updated after each round of dispatching to reflect changes in value due to driver's state transitions in real time. This means that between two reposition actions, the value network updates approximately $C \approx 150$ times, serving as implicit coordination so that the current reposition computations take into account the results of the preceding reposition action. Secondly, within each reposition computation, exploration is enhanced, and overreaction mitigated by adding stochasticity. This process involves sampling the action from the value distribution (5.16) that prevents all idle drivers from being sent to a single location with the highest value.

5.9.1.3 Value Ensemble (Grounding)

In a real-time environment, we continue to manage and update V_θ. This is done using the results derived from each dispatching round according to (5.15). To ensure that daily supply-demand dynamics patterns are accounted for, V_θ is periodically 'reinitialized' with the use of an ensemble scheme that weighs the latest state of V_θ and the snapshot of the offline-trained V_{ope}^t, as described in Sect. 5.8.2.

To clarify, let $\mathcal{E}$ denote the set comprising change time points when the re-ensemble (grounding) is triggered. If the current time step is $t \in \mathcal{E}$, re-ensemble takes place as illustrated below:

$$\forall s, \quad V_\theta(s) \leftarrow \omega V_\theta(s) + (1 - \omega) V_{ope}^t(s). \tag{5.17}$$

Here, $\omega > 0$ is a hyperparameter that balances the weighting between the online value function and offline-trained values. V_{ope} is trained with the current timestamp as part of the input, implying that $V_{ope}^t(s)$ can be obtained by fixing time at t for each state s. The set $\mathcal{E}$

can be set by learning a segmentation on the historical aggregated order time series, which will identify the temporal boundaries of the order distributional shift (Truong et al. 2020).

Though similar, V_θ and V_{ope} have nuances when it comes to learning mechanisms. For instance, while the complete driver trajectory is known and available for learning the offline value V_{ope}, for V_θ, only the partial driver trajectory is available due to online, temporally sequential updating. This means that at the time step t, V_{ope}^t reflects the historical trajectories from t until the episode end, but V_θ trains on the trajectories from the beginning of the current episode up to time t. Update (5.17) helps to reflect recurring time-varying patterns across the history of episodes, while considering individual variations of the current episode when combining both V_{ope}^t and V_θ.

The unified framework brings together offline policy evaluation and online on-policy learning for both order dispatching and driver repositioning. The complete approach is captured in Algorithm 5.1 and illustrated in Fig. 5.6. At the beginning of each episode, the state value network V is initialized with random weights θ, and the offline policy network V_{ope} pretrained accordingly, as described in Sect. 5.8, is obtained.

In the main loop, for a dispatching round at time t (usually every two seconds), the value network is re-ensembled within a change time set. If $t \in \mathcal{E}$, this involves reinitializing the value network as per (5.17). For a tabular value function, this is a weighted average

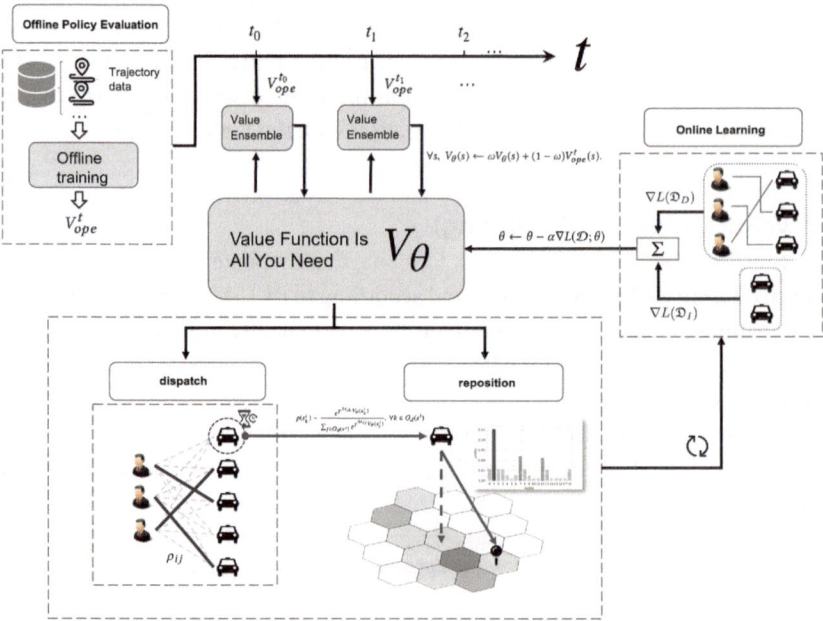

Fig. 5.6 V1D3's illustration. The globally shared value function that is continuously updated by both online learning and the value ensemble from the offline model, creates the backbone of the system. *Source* Tang et al. (2021)

between V_θ and V_{ope}^t, each state in the table is traversed. For networks that represent value functions, this operation can be executed by knowledge distillation (Hinton et al. 2015) with the right-hand side of (5.17) acting as the distillation target. The distillation dataset could be a combination of each state observed so far during the current episode and a sample of historical states.

5.9.1.4 Simulation Results

The V1D3 model outperformed both *first-prize* winners of the dispatch and repositioning tracks in the KDD Cup 2020 RL benchmarks. The superior performance of V1D3 in both order dispatching and driver repositioning was significantly facilitated by the online learning component, which allows V1D3 to swiftly adapt to changes occurring in the online environment.

This claim is supported by the results of further experiments investigating the response of V1D3 to the occurrence of irregular events. The results, presented in Fig. 5.7, provide an empirical overview of how the values of this model change in response to these events. Two cases of sudden changes were examined: the addition of new drivers and the addition of new orders.

In the first case, new drivers, or supplies, suddenly become available in a particular area, usually due to the completion of previous orders or drivers coming online before the start of a large event. In response, V1D3 decreases the values of the grid, dissuading more drivers from being sent to the location. The response is strongest at the center of the grid or radius 0, where the new supplies were created. As time elapses, the additional drivers are dispatched, and the state of the system gradually returns to normal, with the value delta reducing to zero.

In the second case, demand or passenger requests suddenly increase at one location, typically due to the completion of a significant event such as a concert. After such an event, V1D3's value delta increases to attract more drivers to the location and satisfy increased demand. Thereafter, as the new orders are being fulfilled, the state of the system returns to normal, with the value delta decreasing to zero.

Overall, these results demonstrate the capability of V1D3 to quickly adapt to variations in both supply (drivers) and demand (orders) using dynamic value responses. This capability plays a crucial role in balancing the supply and demand of ridesharing markets, improving both driver efficiency and passenger answer rates.

5.9.2   Real-World Deployment

At the time of writing, multiple major ridesharing platforms around the world have deployed online RL algorithms for dispatch in production (Eshkevari et al. 2022; Han et al. 2022; Azagirre et al. 2023). Here, we provide a recount based on Eshkevari et al. (2022) that

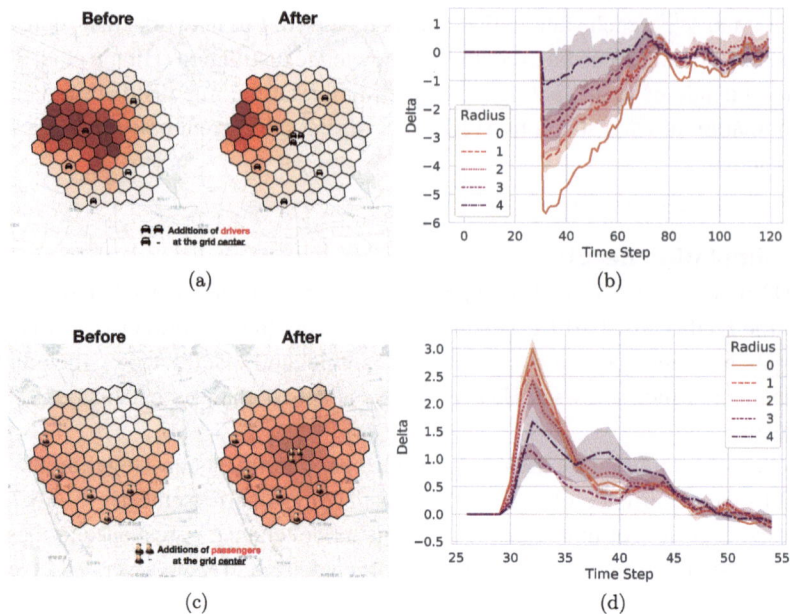

Fig. 5.7 a and **c**: Value distribution before and after additions of new drivers (orders) at the grid center; **b** and **d**: Each color represents the value distributions of the sets of cells at different radius to the center, e.g., radius 0 denotes the grid center. The values decrease (increase) in response to new additional supplies (demands) as desired. The magnitude of the response gradually diminishes as time elapses and as the distance to the grid center increases. Tang et al. (2021)

introduces a few highly effective practical algorithmic enhancements designed for the real-world environment.

5.9.2.1 Expectation-Based Updates

The essence of the online updates from Sect. 5.9.1.1, demonstrated by Eq. 5.18:

$$V_s \leftarrow V_s + \alpha[r_{ss'} + \gamma V_{s'} - V_s], \tag{5.18}$$

suggests a viable update method, as long as the on-policy action secures a transition from s to s'. However, practical applications present a more complex scenario, where, despite the policy proposing an action, the outcome remains uncertain, contingent upon acceptance by both driver and passenger. This necessitates modifications in the update process to integrate the uncertain outcomes of deterministic actions. The revised update rule, as displayed below, handles this problem:

$$V_s \leftarrow V_s + \alpha\left[\sum_{s'} p_{s'}(r_{ss'} + \gamma V_{s'}) - V_s\right] \tag{5.19}$$

$$\sum_{s'} p_{s'}(r_{ss'} + \gamma V_{s'}) = p_c(r_{ss'} + \gamma V_{s'}) + (1 - p_c)(0 + \gamma V_s). \tag{5.20}$$

Equation 5.20 devises a pseudo-sampled value estimate for state s, given two potential outcomes for an on-policy action with a $s \rightarrow s'$ transition: completion or rejection. The completion chance p_c is derived from classification models trained on a vast amount of historical data. The update captures the reward and transaction if the trip is completed in the first term of (5.20), whereas the second term considers the likelihood of cancellation, implying no reward or state transition.

While some may suggest (5.20) can be simply adapted if the value gradient step is post actual agents' transitions, in practicality, this approach introduces deployment challenges regarding system infrastructure. For post-transition update schemes to function, a reliable and low-latency feedback system that reports real-time transition outcomes is necessary. Moreover, since the transitions correspond to trips, the post-transition updates would be much delayed. Due to this, a pre-transition update scheme tends to be more practical and easier to deploy.

5.9.2.2 Reward Smoothing

As the state's value function gets updated based on real-time order requests, each request, linked with a price that is dynamically determined by various algorithms and marketing strategies (e.g., dynamic prices and incentives), contributes to temporary variations. Ideally, the value function should remain unaffected by these temporary alterations to enhance its robustness. Therefore, a reward smoothing procedure is introduced to maneuver the nominal price value and control fluctuations. This process is carried out by the update rule, illustrated in Eq. 5.21:

$$S[grid_o] = \beta \cdot S[grid_o] + (1 - \beta) \cdot price_o. \tag{5.21}$$

In the equation above, β stands for a momentum coefficient, and S is a collection that stores the grid-based smoothed price values. For each order o, $grid_o$ and $price_o$ refer to the starting grid and price, respectively. This mechanism favors the model in reducing variances in the updating signals received by the platform, thereby supporting a similar mechanism discussed in Dong et al. (2021) for controlling the RL training process's stability.

5.9.2.3 Graph Edge Standardization

With respect to RL-based dispatching solutions, the value update term (δ in (5.14)) operates as a utility function in the dispatch graph. Despite the functionality of this mechanism, it lacks the capacity to adjust the relative weights between the components (instant reward and residual value). Moreover, the original function struggles to assign weight factors to penalty terms introduced for business purposes. As a solution, a new utility function for graph edge weight computation comes into play. This function, given in Eq. 5.22, is more flexible and addresses the challenges mentioned:

$$E_{ss'} = p_{ss'} \cdot [w_{rew} \cdot r^*_{ss'} + w_{res} \cdot (\gamma V_{s'} - V_s)^* - w_p \cdot f(ss')^*]. \tag{5.22}$$

Equation 5.23, termed as Stdizer_update, and Eq. 5.24, labeled as Standardize, establish the standardization step:

$$\text{Stdizer\_update:} \quad \begin{cases} m \leftarrow \beta_1 m + (1 - \beta_1)x \\ v \leftarrow \beta_2 v + (1 - \beta_2)(x - m)^2 \end{cases} \tag{5.23}$$

$$\text{Standardize:} \; x^* = Sigmoid((x - m)/\sqrt{v}) \tag{5.24}$$

In this process, any given x (reward, residual value, or penalty) goes through standardization, resulting in x^*, a bounded value within the range of 0 to 1. This facilitates a homogeneous and probabilistic combination of components, making hyperparameter optimization methods like Bayesian Optimization (BO) applicable to tune $w_{\{res,rew,p\}}$.

5.9.2.4 Real-Time Adaptive Graph Pruning

The construction of a bipartite graph from standardized graph edges and the identification of the optimal pair subset by the maximum matching algorithm (Kuhn 1955) could result in low completion probability batches of driver-order assignments. In such instances, constructing graphs from all possible edges would lead to a majority of cancellations, resulting in missed immediate rewards and inefficient driver allocation. To avoid this, the graph can be pruned based on pair probability, which can be achieved through different means, including the use of a fixed probability threshold.

However, this simple approach often fails to handle varying supply-demand levels accurately. For instance, peak hours may require a higher probability threshold due to the complexity of the graph (with many assignment options), whereas quieter ones would not. Therefore, a dynamic threshold adjustment powered by expert knowledge (for example, a nonlinear function proportional to the temporal demand level) or an adaptive mechanism that learns from real-time feedback could be more effective. In this study, the latter is explored, proposing a multi-arm bandit (MAB) solution to address this need.

5.9.2.5 RL in the Wild (RLW)

By combining the algorithmic elements above, a full picture of the online reinforcement learning algorithm is established. The procedural description of the proposed algorithm is presented in Algorithm A.4. Updating the value function using online samples shown in

Eq. 5.18 is analogous to the update rule in stochastic gradient descent (SGD). Therefore, mathematical practices that improve the performance of the vanilla SGD algorithm can be applied in this scenario. In particular, ADAM (Kingma and Ba 2014) is a widely used optimization algorithm that can accelerate the learning process in which two well-known techniques—SGD with momentum and Root Mean Square Propagation (RMSP)—are combined. In the present work, we use ADAM for the gradient ascent updates of the state value functions. Intuitively, the integrated ADAM mechanism can adapt to the variance, magnitude and frequency of value updates.

After the AB test, RLW has been fully deployed in a large pilot city on a major ridesharing platform. The metric is a mixture of the performance measures introduced in this paper. The onset of the treatment- the first day of deployment- is marked with the vertical line. The eval method here is synthetic DID with the control basket covering a rich set of cities within the same geography. This analysis shows that the deployment has a clear positive impact-nearly 5.3% on the dispatching performance in the pilot city. In comparison to the findings from AB testing, the rate of improvement is higher here. This is in part due to the inherent issues in time split tests which could underestimate the long-term effects. On the other hand, this result is quite close to the simulation results reported in Eshkevari et al. (2022).

Vehicle Repositioning

6

6.1 Taxi Routing

Vehicle repositioning from a single-driver perspective (i.e., taxi routing) has a relatively long history of research since taxi service has been in existence long before the emergence of rideshare platforms. Likewise, research on RL-based approaches for this problem (see Table B.4) also appeared earlier than that on system-level vehicle repositioning.

For the taxi routing problem, each driver is naturally an agent, and the objective thus focuses on optimizing individual reward. Common reward definitions include trip fare (Rong et al. 2016), net profit (income—operational cost) (Verma et al. 2017), idle cruising distance (Garg and Ranu 2018), and ratio of trip mileage to idle cruising mileage (Gao et al. 2018). Earlier works (Han et al. 2016; Wen et al. 2017; Verma et al. 2017; Garg and Ranu 2018) optimize the objective within a horizon up to the next successful match (i.e., A-to-B and A-to-C in Fig. 6.1), but it is now more common to consider a long-term horizon, where an episode usually consists of a trajectory over one day (Lin et al. 2018; Shou et al. 2020; Jiao et al. 2021). We illustrate these concepts in Fig. 6.1.

The type of actions of an agent depends on the physical abstraction adopted. A simpler and more common way of representing the spatial world is a grid system, square or hexagonal[1] (Han et al. 2016; Wen et al. 2017; Verma et al. 2017; Gao et al. 2018; Lin et al. 2018; Rong et al. 2016; Jiao et al. 2021; Shou et al. 2020). In this setting, the action space is the set of neighboring cells (often including the current cell). Shou and Di (2020b) explain the justification for this configuration. Determination of the specific destination point is left to a separate process, e.g., pick-up points service (Jiao et al. 2021). The more realistic abstraction is a road network, in which the nodes can be intersections or road segments (Garg and Ranu

---

[1] The square grid system is known as Geohash (https://en.wikipedia.org/wiki/Geohash). The hexagonal grid system (https://h3geo.org/) is more commonly used in practice.

Fig. 6.1 Illustration of the (single-agent) taxi routing problem on a hexagon grid system. The vehicle at its origin position A has the option to reposition to one of the neighboring and current cells. The black arrows represent reposition (idle cruising), and in the two scenarios, the vehicle is matched to a trip request at B and C respectively. The orange arrows represent trip moves, and the orange flags are where the episodes terminate (for long-term horizons)

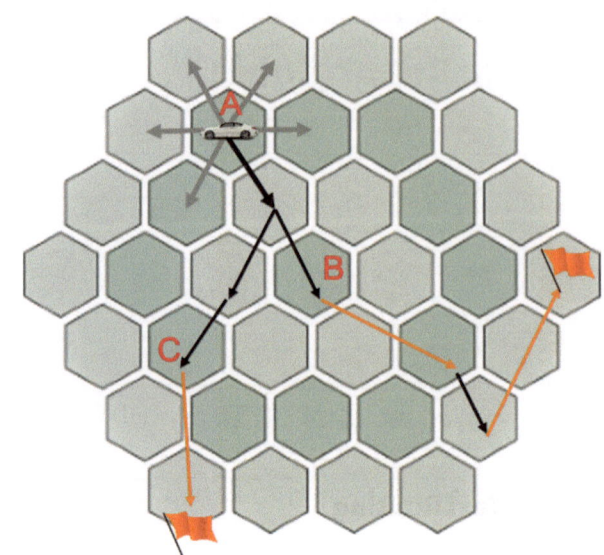

2018; Yu et al. 2019; Zhou et al. 2018; Schmoll and Schubert 2020). The action space is the adjacent nodes or edges of the current node. This approach supports a turn-by-turn guiding policy but requires more map information at run time.

Most of the papers adopt a tabular value function, so the state is necessarily low-dimensional, including spatiotemporal information and sometimes additional categorical statuses. Shou et al. (2020) have a boolean in the state to indicate if the driver is assigned to consecutive requests since its setting allows a driver to be matched before completing a trip request. Rong et al. (2016), Zhou et al. (2018) have the direction from which the driver arrives at the current location. For deep RL-based approaches (Wen et al. 2017; Jiao et al. 2021), richer contextual information, such as SD distributions in the neighborhood, can go into the state.

The learning algorithms are fairly diverse but are all value-based. By estimating the various parameters (e.g., matching probability, passenger destination probability) to compute the transition probabilities, Rong et al. (2016), Yu et al. (2019), Shou et al. (2020), Zhou et al. (2018) adopt a model-based approach and use value iterations to solve the MDP. Shou et al. (2020) further use inverse RL to learn the unit-distance operational cost. Model-free methods are also common, e.g., Monte Carlo learning (Verma et al. 2017), Q-learning (Han et al. 2016; Gao et al. 2018), and DQN (Wen et al. 2017). Jiao et al. (2021) is a hybrid approach in that it performs an action tree search at the online planning stage using estimated matching probabilities and a separately (offline) learned state value network developed in Tang et al. (2019). Garg and Ranu (2018) is in a similar spirit by augmenting the multi-arm bandits with Monte Carlo tree search.

6.2   Case Study: Search-Based Decision-Time Planning

We will dive deep into a taxi routing algorithm developed in Jiao et al. (2021) that performs tree-search for decision-time planning and leverages the driver-level value network (Tang et al. 2019; Qin et al. 2020a) for bootstrapping. This algorithm focuses on tackling the taxi routing problem in the context of ridesharing platforms.

6.2.1   Environment Dynamics

To understand the algorithm, let us start with the environment dynamics of the taxi routing problem. In this problem, the drivers are assigned to trip orders obtained through a ride-hailing platform. The dispatching of orders to drivers takes place in a batch fashion within a time window of a few seconds (Tang et al. 2019; Xu et al. 2018; Zhang et al. 2017). The driver is required to pick up the passenger from the origin location and transport them to the destination. Upon completing the trip, the driver becomes idle and is available for repositioning. If the idle time exceeds a certain threshold (L minutes), the driver performs a repositioning task by cruising to a specific destination. During this repositioning, the driver is still eligible for order assignment. For modeling purposes, it is assumed that the driver can always complete a repositioning task before being matched to an order. However, this assumption does not impact the generality of the formulation, and the policy is not restricted by this assumption in any way. The objective of the algorithm is to maximize the income efficiency (or income rate), which is typically measured as income per online hour (IPH), for individual drivers or groups of drivers.

6.2.2   Semi-MDP Formulation

To formulate the taxi routing problem mathematically, we can model the driver's trajectory as a semi-Markov Decision Process (Semi-MDP) (Tang et al. 2019; Qin et al. 2020a). The Semi-MDP is defined by the following components:

State The driver's state s includes spatiotemporal information about the driver's location l and time t. Additionally, it can include additional supply-demand contextual features f. The state is denoted as $s = (l, t, f)$, where l represents the location, t represents the time, and f represents additional contextual features. The state can be further divided into basic spatiotemporal features (l, t) or additional supply-demand features f.

Option The eligible actions for the driver include vehicle repositioning and order fulfillment. These actions are temporally extended and are called options in the context of Semi-MDP. A basic repositioning option for the driver is to move towards a destination in one of the neighboring cells defined by a hexagonal grid system. The order fulfillment option represents the driver accepting and completing an order. The duration of an option is denoted as τ_o.

Reward The reward associated with each option is determined by the price of the trip or the cost of repositioning. The reward for repositioning is denoted as $c_o \leq 0$, while the reward for order fulfillment is $p_o > 0$. The immediate reward of a transition is denoted by r, where $r = c_o$ for repositioning and $r = p_o$ for order fulfillment. The estimated versions of the duration, price, and cost are denoted as $\hat{\tau}_o$, $\hat{p}_o$, and $\hat{c}_o$ respectively.

Transition The transition dynamics for the driver depends on the state and the selected option. The transition given a repositioning option is deterministic, while the transition probability given an order fulfillment option $P(s'|s, o_d)$ is the probability of the driver transitioning to state s' given the current state s and the dispatching option o_d.

The objective of the algorithm is to learn a policy that maximizes the cumulative income rate (IPH). The IPH is defined as the ratio of the total price of trips completed during an episode to the total online hours logged by an individual driver or a group of drivers. The individual-level IPH for a driver x is denoted as $p(x)$, and the group-level IPH for a driver group X is denoted as $p(X)$.

6.2.3   Learning State Values

To formulate the value-based policy search algorithm (VPS), we first need to learn the state values associated with the semi-MDP. In this section, we will describe the learning of state values using a method called Dual Policy Evaluation (DPE).

The state values $V(s)$ can be decomposed into two components based on whether the driver is being dispatched or not:

$$V(s) = p_d^{(s)} V(s|dispatch) + p_{id}^{(s)} V(s|idle), \tag{6.1}$$

where $V(s|dispatch)$ and $V(s|idle)$ are the conditional value functions given the driver being dispatched or idle at state s, respectively. The dispatch probabilities $p_d^{(s)}$ are estimated separately.

DPE is a method that jointly learns $V(s|dispatch)$ and $V(s)$ while reducing the stochasticity in the policy to prevent increased variance. DPE updates the conditional value function $V(s|dispatch)$ based on its expected value $V(s')$, rather than the next-state value function $V(s'|dispatch)$, which reduces the variance in the updates. The update equations for DPE can be formulated as follows:

$$V(s_0|b) \leftarrow \frac{R_b(\gamma^k - 1)}{k(\gamma - 1)} + \gamma^k V(s_k), \tag{6.2}$$

where $V(s_0|b)$ is the conditional value function for the initial state s_0 and option b. R_b is the reward obtained from the option b, γ is the discount factor, and k is the number of transition steps. DPE updates the conditional value functions iteratively until convergence.

DPE utilizes a neural network to represent the value function, with separate branches for $V(s|dispatch)$ and $V(s)$. The neural network is trained using historical trajectories and variance reduction techniques to improve convergence speed and stability. The policy is updated using the conditional value functions learned from the trajectory data.

6.2.4 Value-Based Policy Search (VPS)

VPS is a planning algorithm that selects the optimal action using decision-time planning. Given the learned value functions $V(s)$ and $V(s|dispatch)$, the algorithm calculates the value of each action-option pair using an environment model and selects the action with the highest value. The algorithm can be seen as a path-value approximation approach, where the value of n-step look-ahead repositioning paths is estimated and the optimal path is selected based on their values.

VPS can be implemented using a breadth-first search method to generate all paths of a certain length from the current grid cell. The algorithm then calculates the value of each path and selects the first step of the path with the maximum value as the repositioning action. The state value networks are used to estimate the values of the paths, and the algorithm iteratively updates the state value networks based on the new data generated by the current policy.

Figure 6.2 provides a visual representation of the planning and bootstrapping process in the VPS algorithm. The algorithm uses decision-time planning to compute the state-option values for each possible repositioning action at each decision point. The values are then used

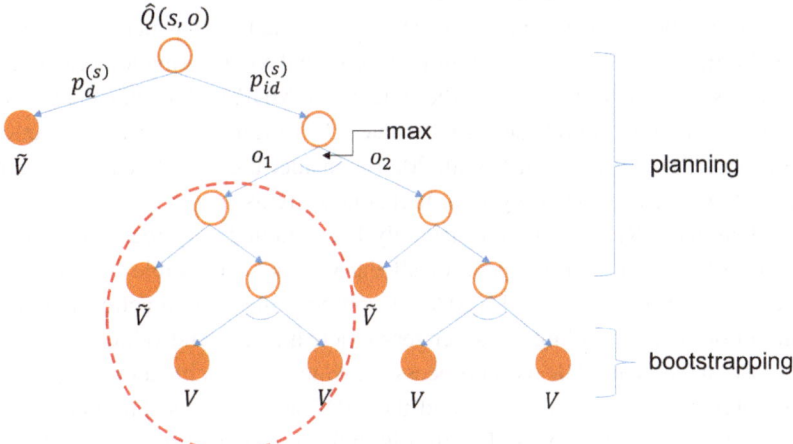

Fig. 6.2 Planning + bootstrapping perspective of the proposed algorithm. At the leaf nodes, bootstrapping is used through the given state-value function V. The subtree within the red loop illustrates the back-up of values. *Source* Jiao et al. (2021)

to select the best repositioning action at each decision point. The state value networks are updated using the new trajectory data generated by the current policy, which improves the accuracy of the value estimations and the overall decision-making process.

VPS provides an efficient method for decision-time planning in the taxi routing domain. By utilizing the learned state values and the environment model, the algorithm can make informed decisions on repositioning actions, maximizing the income efficiency of the drivers. The algorithm combines the strengths of reinforcement learning and tree-search algorithms to tackle the complex single-vehicle repositioning problem in the context of ride-hailing.

6.3   System-Level Repositioning

The problem formulation most relevant to the ridesharing service provider is system-level vehicle repositioning. Similar to order matching, the ridesharing platform reviews the vehicles' states at fixed time intervals which are significantly longer than those for order matching. Idle vehicles that meet certain conditions, e.g., being idle for sufficiently long time and not in the process of an existing reposition task, are sent reposition recommendations, which specify the desired destinations and the associated time windows. The motivation here is to explicitly modify the current distribution of the available vehicles so that collectively they are better positioned to fulfill more requests more efficiently in the future. Figure 6.3 explains the idea with a concrete example. If the vehicles reposition independently (following the orange arrows), they both move to the orange-circled area and there will be a surplus of supply while the demand in the green-circled area will not be served. In contrast, if the vehicles coordinate and the one in the south repositions by the blue arrow, both vehicles will be matched, and all the requests are served.

The agent can be either the platform or a vehicle, latter of which calls for a MARL approach. All the works in this formulation have global SD information (each vehicle and request's status or SD distributions) in the state of the MDP, and a vehicle agent will additionally have its spatiotemporal status in the state. The rewards are mostly the same as in the taxi routing case, except that (Mao et al. 2020) consider the monetized passenger waiting time. The actions are all based on grid or taxi zone systems.

The system-agent RL formulation has only been studied very recently, in view of the intractability of the joint action space of all the vehicles (see Table B.5). To tackle this challenge of scalability, Feng et al. (2020) decompose the system action into a sequence of atomic actions corresponding to passenger-vehicle matches and vehicle repositions. The MDP encloses a 'sequential decision process' in which all feasible atomic actions are executed to represent one system action, and the MDP advances its state upon complete of the system action. They develop a PPO algorithm for the augmented MDP to determine the sequence of the atomic actions. The system policy in Mao et al. (2020) produces a reposition plan that specifies the number of vehicles to relocate from zone i to j so that the action space is independent from the number of agents (at the expense of additional work at execution).

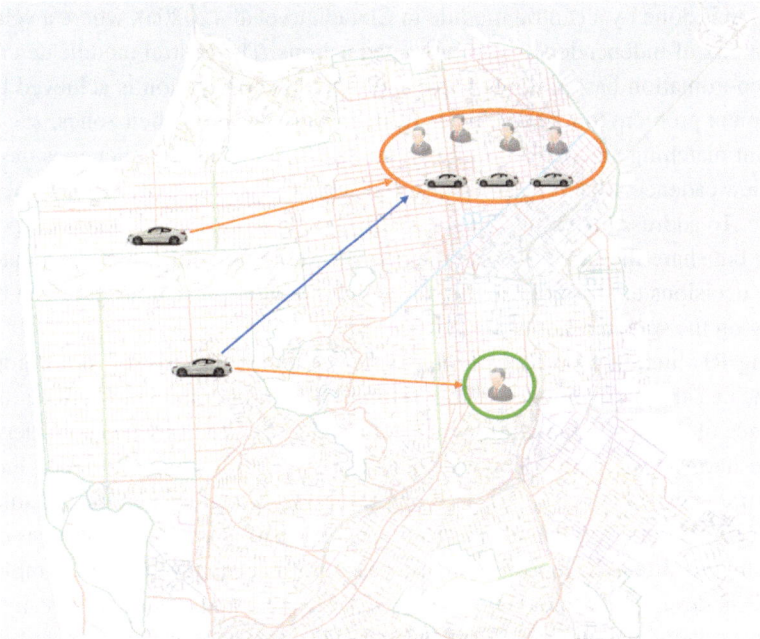

Fig. 6.3 Illustration of system-level vehicle repositioning. The requests in the orange-circled and green-circled areas appear in the future w.r.t. the time of repositioning. The empty vehicles are existing ones in the orange-circled area. The orange and blue arrows represent potential reposition moves

The agent network, trained by a batch AC method, outputs a value for each OD pair, which after normalization gives the percentage of vehicles from each zone to a feasible destination.

The vehicle-agent approaches have to address the coordination issue among the agents. Lin et al. (2018) develop contextual DQN and AC methods, in which coordination is achieved by masking the action space based on the state context and splitting the reward accrued in a grid cell among the multiple agents within the same cell. Oda and Joe-Wong (2018) treat the global state in grid as image input and develop an independent DQN method. They argue that independent learning, equipped with global state information, works quite well compared to an MPC-based approach. The zone structure in Liu et al. (2020) is constructed by clustering a road-connectivity graph. A single vehicle agent is trained with contextual deep RL and generates sequential actions for the vehicles. Zhang et al. (2020b) also train a single DQN agent for all agents, but with global KL distance between the SD distributions similar to Zhou et al. (2019). The DQN agent is put in tandem with QRewriter, another agent with a Q-table value function that converts the output of DQN to an improved action. Shou and Di (2020b) approach the MARL problem with bilevel optimization: The bottom level is a mean-field AC method (Li et al. 2019) with the reward function coming from a platform reward design mechanism, which is tuned by the top level Bayesian optimization. Agent coordination is done by a central module in Chaudhari et al. (2020a), where a vehicle agent executes a mix of independent and coordinated actions. The central module determines the need for coordination based on SD gaps, and explicit coordination is achieved by solving an assignment problem to move vehicles from excess zones to deficit zones.

For joint matching and repositioning optimization, one major challenge is the heterogeneous review cadence. Matching and reposition decisions are typically made asynchronously in practice. To address this issue, Tang et al. (2021) allow the two modules to operate independently but share the same spatiotemporal state value function which is updated online. If the two decisions are formulated into the same problem, the action space can be masked depending on the state (Holler et al. 2019).

Existing RL literature on repositioning often assumes the drivers' full compliance to reposition, i.e., the autonomous vehicle setting. How non-compliance affects the overall performance of a reposition algorithm is a natural question to ask when considering a real-world ridesharing system, in which we expect to see a combination of drivers' independent cruising strategies (Urata et al. 2021; Wong et al. 2014) and system-guided idle cruising behavior. It is also interesting and practically necessary to investigate incentives design and strategies that facilitate the repositioning process. In Zhu et al. (2021), for example, a mean-field MDP is developed for modeling drivers' strategies, and empirical investigations are performed on how spatiotemporal driver incentives affect driver behavior and the system performance.

6.4   Case Study: Large Fleet Management via SARSA

We describe in more details about a SARSA-based algorithm tailored for the (repositioning) management of a large fleet (Jiao et al. 2021). The scale of the fleets necessitates global coordination strategies to prevent causing supply-demand imbalances. For instance, under an independent strategy, all vehicles sharing the same spatial-temporal state may veer towards what might seem like an attractive high-value destination, leaving other areas undersupplied—an 'over-reaction' phenomenon. Learning algorithms must then be tweaked to better integrate group coordination into the repositioning policy.

Our objective is to maximize the cumulative IPH, which is the ratio of the total price of the trips completed during an episode and the total online hours logged by a driver (individual level) or a group of drivers (group level). The individual-level IPH for a driver x is defined as

$$p(x) := \frac{c(x)}{h(x)}, \tag{6.3}$$

where $c(\cdot)$ is the total income of the driver over the course of an episode, and $h(\cdot)$ is the total online hours of the driver. Similarly, the group-level IPH for a driver group X is defined as

$$p(X) := \frac{\sum_{x \in X} c(x)}{\sum_{x \in X} h(x)}. \tag{6.4}$$

In the context of large-fleet repositioning, a single action-value network is used to generate policies for each managed vehicle, mirroring the approach outlined for smaller fleets. This approach has several merits:

- It is indifferent to the ever-changing number of agent vehicles, providing a natural solution to account for drivers going on and off duty. Learning value functions specific to each agent—as is typical of numerous multi-agent reinforcement learning (MARL) methods— proves problematic in this context.
- It avails a diverse range of training data and enhances offline reinforcement learning.
- Operationally, a singular value network simplifies the training process, making it suitable for vehicle repositioning—a high usage service.

In this scenario, we are motivated to directly learn the action-value function, $Q(s, o)$. Reasons for this include the straightforward integration of algorithmic elements of MARL, the convenience of incorporating real-time contextual features, and fast speeds. We designed our methodology to meet strict latency requirements for online planning, making rapid decision-time planning crucial. Comparatively, an action-value network significantly outperforms Value-based Policy Search (VPS, see Sect. 6.2) because it eliminates the need for tree search and supports batch inference. To cite a case in point, a single action-value network demonstrates a QPS (queries-per-second) capacity approximately six times higher than VPS. In the context of managing a large fleet, this is a crucial determinant.

We use deep SARSA to learn the state-action value function in a similar mode as the CVNet. It is clear in this context that vanilla deep SARSA is merely VPS with an expansion depth of one. If we expand $Q(s, o)$ and presuppose a stochastic policy π, we find:

$$Q^\pi(s, o) = E_{o' \sim \pi(\cdot | s')} \left[r(s, o) + \gamma Q^\pi(s', o') \right]$$
$$= r + \gamma V^\pi(s'),$$

where s' refers to the state that follows s with the option o.

Although the form of deep SARSA is simpler, it allows for the natural incorporation of some very desirable algorithmic elements which we describe in the subsections that follow.

6.4.1   Supply-Demand Context

The repositioning policy of the vehicles is improved by enlarging the state space with additional supply-demand (SD) contextual features in the neighborhood of the agent. These typically include the number of orders and idle drivers awaiting matching. These features enable a more precise representation of the vehicle's state and its surrounding environment, allowing for greater responsiveness to changes. In this context, we include SD features from the six directly adjacent hexagonal grid cells, along with those from the current cell. The adopted global attention mechanism has the form $\alpha_i = \text{softmax}(sd_0^\top W_\alpha sd_i)$, where $i \in \mathbb{Z}, i = [1, \ldots, 6]$, and W_α is a trainable weight matrix in the attention layer. The attention layer assigns scores to each pair of SD features (current cell and one neighboring cell), and these scores are then used to re-weight the neighboring SD vectors through α_i's, creating a dense, reliable context representation vector.

6.4.2   Stochastic Policy

A deterministic policy results in vehicles in the same state being repositioned to the same destination. This is not problematic when the fleet size is small relative to the general vehicle pool, as the likelihood of multiple vehicles being in the same state is low. However, as the scale of the fleet increases, such encounters become more frequent, and the impact of the 'over-reaction' phenomenon becomes more prominent. To alleviate this problem, we introduce a measure of randomness into the repositioning action output by appending a softmax layer to obtain the Boltzmann distribution:

$$\sigma(\mathbf{q})_k = \frac{\exp(q_k)}{\sum_j \exp(q_j)}, \forall k \in \mathcal{K}, \tag{6.5}$$

where $\mathbf{q}$ represents the vector of reposition action-values output, and $\mathcal{K}$ is the set of eligible destinations. While this is a standard approach in reinforcement learning to transform action-values into action probabilities for a stochastic policy, it is particularly useful in cases of vehicle repositioning because it has the desirable property that the reposition allocation of a group of vehicles in the same state aligns well with the action-values.

6.4.3   SD Regularization

When examined from the perspective of a single vehicle, the Semi-MDP does not account for the current supply-demand gap at the destination. To regularize the action values and ensure that the system's repositioning actions consider supply and demand, the action values are penalized based on their respective destination SD gaps:

$$q_k' := q_k + \alpha g_k, \forall k \in \mathcal{K}, \tag{6.6}$$

where g_k is the SD gap of destination k at the decision time.

6.4.4   Real-World Implementation

We developed an AI Driver Assistant application (DA), equipped with our repositioning algorithm. The DA was used in a pilot program for real-world testing on a major ridesharing platform to evaluate its performance against human repositioning strategies. The challenge lies in competing with human drivers who have garnered domain knowledge from experience and are privy to real-time demand information provided by the app.

Deployment testing of the DA with human drivers required considering how recommendations were delivered and how willing drivers were to follow such recommendations. To encourage driver participation, we designed a hierarchical incentive program to ensure that the drivers are online for a significant period and they follow the repositioning recommendations as closely as possible.

AI Driver Assistant

Repositioning recommendations from the DA were delivered through pop-up message cards within the mobile app. Every time repositioning was triggered, a message card appeared on the driver's app, providing repositioning instructions including the destination and the target arrival time. Upon acknowledging the task, GPS navigation was launched to guide the driver to the destination. The system then automatically determined whether the driver had arrived at the destination within the required timeframe. Figure 6.4 shows a set of illustrative screenshots for such an AI driver assistant.

Evaluation Methodology

To evaluate the DA's performance on larger driver groups, we recruited approximately 1200 drivers (300 each for cities K and F, and 600 for city M) from three different cities on the ridesharing platform to participate in our pilot programs.

An important consideration in comparing group level performance is the variability due to different samples of groups. In our case, we performed significance tests by bootstrapping. This process involved treating the experiment group as one sample X, which gave the group

Fig. 6.4 Illustrative screenshots of a driver assistant application

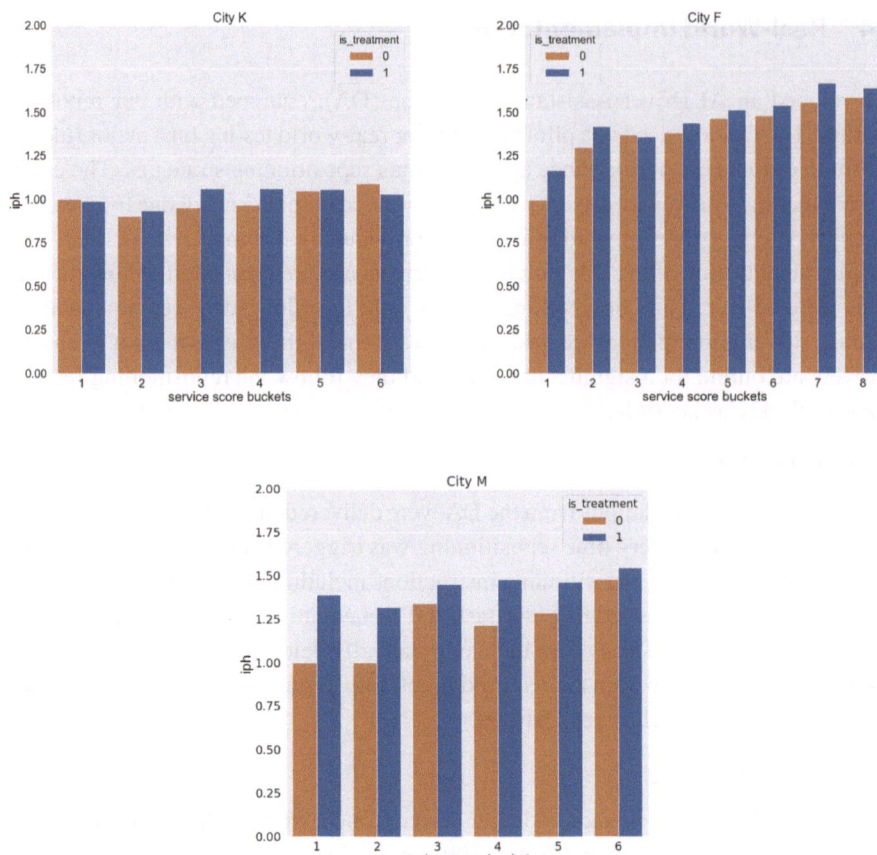

Fig. 6.5 Income efficiency comparison by service score buckets

income rate $p(X)$ as in (6.4). We sampled with replacement to select N control groups of size $|X|$ from a control pool whose size was several times larger than $|X|$.

Results

The evaluation demonstrated significant improvement in both the aggregated IPH and utilization rates for the experiment groups compared to those still using human expert strategies (see Fig. 6.5 and discussion in Jiao et al. (2021))—this suggests that applying our algorithm and deploying the DA substantially enhances the participating drivers' welfare and the platform's overall efficiency.

Routing

Routing is an ambiguous terminology in the context of ridesharing. Routing, which we discuss in Sect. 7.1, commonly refers to low-level navigation decisions on a road network, typically with output of matching and repositioning algorithms as input. This is not to be confused with the *vehicle routing problem*, which is a separate class of macro-level problems that we elaborate in Sect. 7.2.

7.1   Route Guidance (Navigation)

The road network, combined with traffic conditions on the links (exhibited as link costs), forms the traffic network which is a non-stationary stochastic network (Mao and Shen 2018). It is known that standard static shortest-path algorithms do not find the path with minimum expected cost in this case, and the optimal route is not a simple route but a policy (Hall 1986; Kim et al. 2005). There are two types of set-up for the routing problem, depending on the decision review time. In the first type of set-up, each vehicle on the road network selects a route for a given OD pair from a set of feasible routes, e.g., as shown in Fig. 7.1. The decision is only reviewed and revised after a trip is completed. Hence, it is called route planning or route choice. When the routes for all the vehicles are planned together, it is equivalent to assigning the vehicles to each link in the network, and hence, the problem is called traffic assignment problem (TAP), which is typically for transportation planning purposes. In the second type of set-up, the routing decision is made at each intersection to select the next outbound road (link) to enter. These are real-time adaptive navigation decisions for vehicles to react to the changing traffic state of the road network. (See Fig. 7.2 for an illustration.) The problem corresponding to this set-up is called dynamic routing, dynamic route choice, or route guidance.

© The Author(s), under exclusive license to Springer Nature Switzerland AG 2025   61
Z. (Tony). Qin et al., *Reinforcement Learning in the Ridesharing Marketplace*,
Synthesis Lectures on Learning, Networks, and Algorithms,
https://doi.org/10.1007/978-3-031-59640-7_7

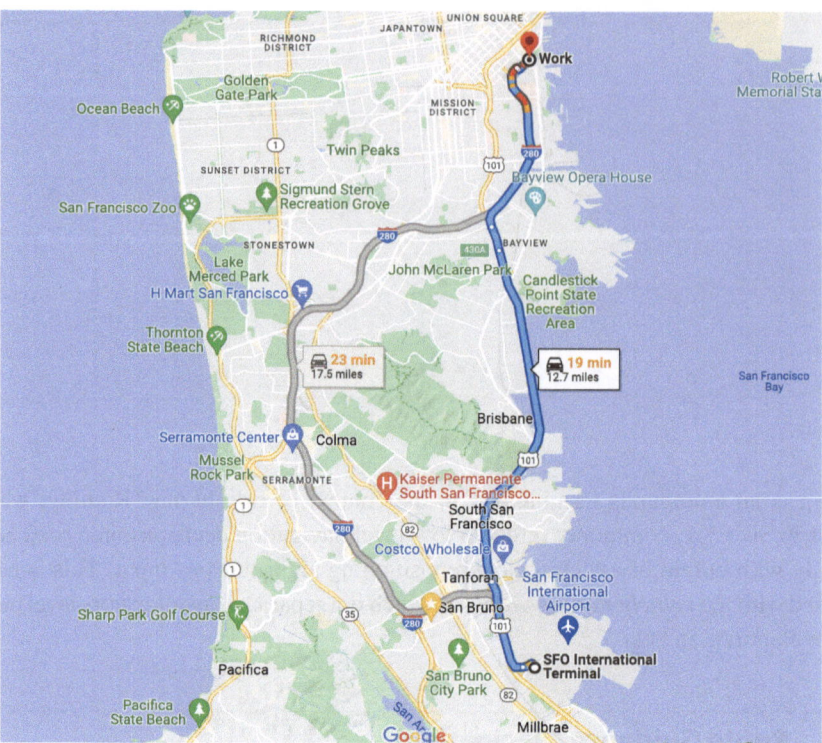

Fig. 7.1 Two planned routes at a given time of day between SFO international airport and downtown San Francisco. *Source* Google Maps

7.1.1 Traffic Assignment Problem (Route Planning)

Routing on a road network is a typical multi-agent problem, where the decisions made by one agent has influence on the other agents' performance, simply in that the congestion level of a link depends directly on the number of vehicles passing through that link at a given time and has direct impact on the travel time for all the vehicles on that link within the same time interval. The literature for route planning and TAP often consider the equilibrium property of the algorithms when a population of vehicles adopt them. TAP is typically from a traffic manager's (i.e., system's) perspective. Its goal is to reach system equilibrium (SE, or also often referred to as the system optimum). Some works focus on route planning or TAP from an individual driver's perspective and maximize individual reward. These algorithms try to reach user equilibrium (UE) or Nash equilibrium, under which no agent has the incentive to change its policy because doing so will not achieve higher utility. This is the best that selfish agents can achieve but may not be optimal for the system.

Value-based RL is by far the most common approach for route planning and TAP aiming to reach UE (see Table B.6). In the MDP formulation, the agent is a vehicle (or equivalently,

Fig. 7.2 Turn-by-turn routing (navigation) for a given pair of origin and destination. *Source* Google Maps

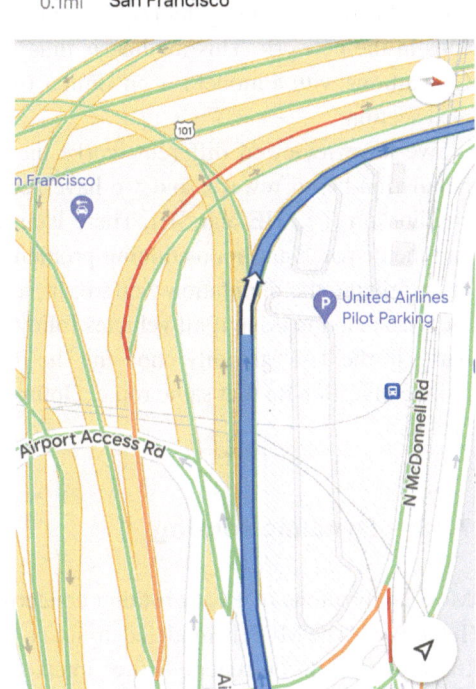

a task) with a given OD pair. The objective is to minimize the total travel time for an individual vehicle (task) (Mainali et al. 2008; Ramos et al. 2018; Zhou et al. 2020a), i.e., the agent is selfish. The immediate reward is the total travel time of a trip for an individual and a particular run. This MDP is stateless, so strictly speaking, it is a multi-arm bandits or contextual bandits problem (Li et al. 2010) if considering time as a contextual feature. The action to take at decision time is to select a route from the set of feasible routes for the associated OD pair (Ramos et al. 2018; Zhou et al. 2020a; Bazzan and Chira 2015). Figure 7.1 illustrates two feasible routes for a given OD pair from SFO airport to downtown San Francisco. The value function that governs the route choice decisions represents the long-term expected travel time for the trip identified by the given OD pair. Due to the multi-agent nature, the environment w.r.t. each agent is non-stationary in that the reward function is changing with the policy updates from the other agents. Empirical convergence to UE is demonstrated by Ramos et al. (2018). Zhou et al. (2020a) further develop a Bush-Mosteller RL scheme for MARL and formally establishes its UE convergence property.

We also highlight some unique features of the papers. Ramos et al. (2018) consider a different objective from the common and minimizes the driver's regret. To do that, the Q-learning updates are modified using the estimated action regret, which can be computed by local observations and global travel time information communicated by an app. Bazzan and Chira (2015) propose a hybrid method, with Q-learning for individual agents and Genetic Algorithm for reaching system equilibrium, minimizing the average travel time over different trips in the network. This method is thus able to achieve SE. Mainali et al. (2008) adopt Q-iterations with a model set-up similar to that of dynamic routing to be discussed in the next section.

We have reviewed different works that leverage local independent reward or global population-based reward, and we have seen their respective association to the nature of equilibrium (i.e., UE and SE). There is an interesting analogy to draw between the TAP and the dispatch and repositioning problems in Chaps. 5 and 6 in terms of the challenges in optimizing the population welfare. In vehicle repositioning, for example, we encounter the over-reaction issue if all vehicles follow an independently learned value function. Similarly, in the TAP, greedily choosing the route of the best independent value will lead to more congestion on that same route. Optimizing the global reward is intrinsically a MARL problem.

7.1.2 Dynamic Routing

Most applications of RL to routing concern with the *dynamic routing* (DR) problem (see Table B.6). The MDP is modeled around a vehicle agent. The basic state information is the traffic state of the current node (i.e., intersection). Some works consider state features of the neighboring nodes (Kim et al. 2005; Mao and Shen 2018) so that the agent has a broader view of the environment. The action space comprises the set of outbound links (i.e., roads) or adjacent nodes from the current node, so the policy provides a turn-by-turn navigation guidance until the destination is reached. While it is most common to use travel time on a link as the reward function, Tumer et al. (2008), Grunitzki et al. (2014) stand out by defining a new form called difference reward, which is the difference in average travel time on a link with and without the agent in the system. This applied to only a reward function dependent on the number of agents using the traversed link. In particular, travel distance cannot be used to define a difference reward. Whether solving a specific formulation achieves UE or SE depends on the reward function used. The average travel time on a link is a global reward because it is an aggregate of local rewards (i.e., individual travel times) of all the agents on that link. The difference reward, by definition, is a global reward that also reflects individual effect. If all the agents in the system learn by global reward (Tumer et al. 2008; Grunitzki et al. 2014; Shou and Di 2020a), then the system is expected to achieve SE. Otherwise, the agents learn by their local rewards, and we will have UE or Nash equilibrium (Kim et al. 2005; Yu et al. 2012; Mao and Shen 2018; Bazzan and Grunitzki 2016; Wen et al. 2019).

Most works in the literature adopt Q-learning or its variant as the training algorithm. We report several notable developments. To tackle the sample efficiency issue of online model-free methods, Mao and Shen (2018) propose an offline batch RL approach (fitted

Q-iterations) with a tree-based function approximator (Extreme Randomized Trees) that empirically shows good convergence property. Hierarchical methods have also been adopted to address the complexity of a large-scale problem. In Wen et al. (2019), the global road network is divided into sub-networks by differential evolution-based clustering. The top-level network contains only the boundary nodes of the original network. The top-level policy produces the destination node for a sub-network. The sub-level policy provides link-level guidance to reach its sub-destination. Shou and Di (2020b) adopt a bilevel optimization scheme. At the lower level, a mean-field MARL algorithm solves for the dynamic routing problem for the travelers, while at the upper level, a Bayesian optimization module optimizes the control (i.e., reward parameter of the travelers) by the city planner.

The dynamic nature of the DR problem determines that the ability to effectively leverage real-time traffic data will play a key role in the future. This calls for the necessary data, hardware, and software infrastructure that makes this kind of state and reward information available to the vehicles in the traffic network. There are primarily two possibilities: centrally connected vehicles which receive real-time data from the central command center, or locally connected vehicles which collect and share neighborhood traffic data with each other. The latter is generally more advantageous in practicality and scalability. The centralized training decentralized execution paradigm (see Sect. 5.4) potentially fits well to this scenario.

7.2 Vehicle Routing Problem (VRP)

In contrast to the routing problems in the previous section, VRP concerns with higher level decisions on where a vehicle should go. The road network is often abstracted in this case. VRP has close connection to the ridesharing problem in that variants of VRP could serve as a subroutine of the ride-pooling problem and could even used to model the entire ridesharing problem itself. The main challenge, in the context of ridesharing, is that new demand (a pair of pick-up and drop-off locations) appears in an online nature and has to be inserted into the existing route dynamically. So reviewing the RL literature for VRP is not only for the completeness of this book but also essential for one to appreciate the complexity and challenges in tackling ridesharing via RL.

7.2.1 Problem Variants

VRP has many variants in its rich literature, so it is important to be clear on their differences and on the variant that each paper claims to solve. The basic setup of a VRP consists of a transportation network $\mathcal{G} := (\mathcal{V}, \mathcal{E}, w)$ and a fleet of K vehicles, where $\mathcal{V}$ is the set of nodes (customer and depot locations), and $\mathcal{E}$ is the set of edges such that $e_{ij} \in \mathcal{E}$ indicates that it is possible to travel from node v_i to node v_j. $w(x_{ij}) := w_{ij}$ is the edge cost, typically distance or travel time. The depot $v_0 \in \mathcal{V}$ is a special node where all the vehicles depart

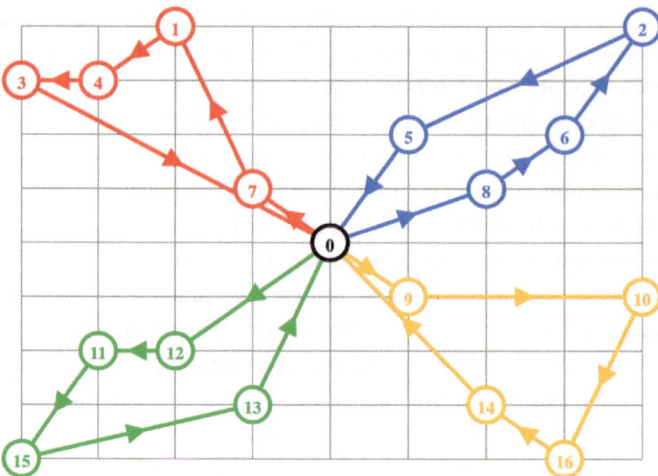

Fig. 7.3 A solution to a vanilla VRP problem for four vehicles or a single vehicle with four trips. Node v_0 is the depot. Each color represents a separate trip. *Source* https://developers.google.com/optimization/routing/vrp

and return at. The vanilla VRP is to find an optimal set of disjoint routes (one for each vehicle) that start and end at the depot, that collectively cover all the nodes, and whose total cost (summing over all the edges in those routes) is minimized. See Fig. 7.3 for a sample solution. A *traveling salesman problem* (TSP) is a special (simplified) instance of VRP, in which there is no depot, and the fleet consists of a single vehicle. In a *capacitated* VRP (CVRP), each node has a demand quantity to be fulfilled by one of the vehicles. Each vehicle has a limited capacity, and it starts from the depot with a full load of goods to fulfill the demand of the nodes on its route. The total capacities of the fleet is sufficiently large, and all the demand has to be fulfilled. A CVRP with *split delivery* allows the demand of each node to be fulfilled by multiple vehicles. In a VRP with *time windows*, each node has a delivery window within which the node has to be visited or its demand has to be satisfied. If the time window constraints are *soft*, they can be violated with the price of a penalty that contributes to the total cost function. In some variants of the VRP, the goods for the demand of a node (destination) has to be picked up from another designated non-depot location (origin) before being delivered to it. This is known as a VRP with *pick-up and delivery*, also known as the *dial-a-ride* problem (DARP). If the fleet consists of electric vehicles (EV), the set of nodes also include charging stations, and each EV has a limited battery capacity, before the depletion of which the EV has to reach a charging station to recharge. In practical situations, a VRP or variant can be *stochastic* and *dynamic* (SDVRP), i.e., its parameters (e.g., demand and travel time) are uncertain, and the requests are not known at the beginning but are revealed sequentially throughout the problem period.

The goal of this chapter is not to provide a complete survey of the VRP literature but rather to point out the representative or unique works that adopt RL to solve VRPs (see Table B.8). A recent review of RL-based methods for solving stochastic dynamic VRP can be found in Hildebrandt et al. (2023) and a more general one in Ulmer et al. (2020).

7.2.2 Single-Vehicle Versus Multi-vehicle Problems

CVRP may appear in different forms, and sometimes the subtle differences may not be stated clearly. Most papers solve the single-vehicle problem where there is only one active vehicle at any time. In the capacitated single vehicle problem, the vehicle can make multiple tours (i.e., passing through the depot multiple times) to fulfill all the demand, but the number of tours is not set in advance. In the multi-vehicle case, a fleet of K vehicles are active simultaneously. For static VRPs, if the number of tours in the single-vehicle case is fixed, then it is equivalent to the multi-vehicle counterpart by treating each tour as a separate vehicle. (For problems with time windows, this equivalence can be achieved by resetting the clock every time a new tour starts.) Otherwise, they are not equivalent in general because the number of tours in the optimal solution for the single-vehicle problem may not be N. For dynamic problems where the requests are not all known a-priori, it is not possible to generate the fixed number of tours in sequence, since one cannot insert a new request to a previous tour. In this case, equivalence can only be achieved by keeping each tour on the same clock and updating the routes with the newly appeared requests at each time step. As we will see below, this would render essentially a multi-vehicle algorithm.

7.2.3 RL Models

The majority of the RL-based methods for VRP models the agent as a vehicle with the system-state visibility. The state thus consists of two types of information: the vehicle state, which includes the vehicle's current location and remaining capacity (for pick-up and delivery, e.g., Ulmer et al. 2020; James et al. 2019; Joe and Lau 2020) or inventory (for homogeneous goods delivery, e.g., Nazari et al. 2018; Kool et al. 2018; Delarue et al. 2020); the system state, which contains the locations of the customer nodes, the demand at each node, and the unserved customers. For pick-up and delivery problems, the system state instead contains the pick-up and delivery locations of the orders. In the case of EVs, the vehicle state additionally contains the vehicle's battery level, and the system state also includes the locations of the charging stations and the number of vehicles available (not in charging). The action of the agent is to specify the next stop (pick-up/delivery location or charging station in the case EV) to visit for the current vehicle. The sequence of actions form a route for the vehicle. When multiple routes/tours are required, the different routes are separated by the insertion of the depot (Nazari et al. 2018; Duan et al. 2020; Kool et al. 2018; Lin et al. 2021). For (dynamic) multi-vehicle problem where decisions for all the vehicles are made at each time step, the agents would generate their actions sequentially to avoid conflicting actions (James et al. 2019; Zhang et al. 2020a). Since the objective of VRPs is typically to minimize total travel distance, the reward is naturally defined as the negative travel distance. For problems with (soft) time window constraints, the negative penalty for constraint violation is added to the reward.

Typically, these methods adopt an encoder-decoder agent network architecture. The encoder is responsible for encoding part or all of the state information into an embedding vector (or context), which, potentially with additional input state features, is fed into the decoder to generate the action one at a time. Bello et al. (2016) develop an policy network based on the pointer network (Vinyals et al. 2015), which consists of an RNN encoder and an RNN decoder. The major novelty over a sequence-to-sequence architecture is that the decoder uses attention mechanism to attend over the embeddings of the input nodes to generate the probability distribution over the input space, thus eliminating the distance disparity in the output with respect to the input, a feature that is particularly suitable for solving TSP and VRP. To reduce the complexity of the encoder and avoid imposing a sequence on the input state features (e.g., customer locations), which is unnecessary in routing problems, Nazari et al. (2018) modify the pointer network with a non-sequential encoder which simply embeds each individual input node. They incorporate the policy network into an AC method and validate the design on a CVRP with split delivery. A few more recent works have adopted this network structure. James et al. (2019) use structural graph embedding (Struct2Vec) for the encoder, since their agent's state additionally contains a vehicle tour graph. In Lin et al. (2021), the encoder has 1D convolution and graph embedding for the input nodes, followed by an attention layer. Duan et al. (2020) include edge features in the state besides the node features of the transportation network. Their encoder is based on graph convolution network with both node and edge inputs. Another work with significant novelty is Kool et al. (2018), which develops a policy network with a transformer-based encoder and a self-attention-based decoder to use in a PG method (REINFORCE) with the baseline computed from deterministic greedy rollout. This training framework has also been adopted by Zhang et al. (2020a) for multi-vehicle VRP with soft time windows, Lin et al. (2021) for EV VRP with time windows, and Duan et al. (2020), which jointly train an MLP-based binary classifier on edge encoding with the policy network output as labels. They have tested their method on a CVRP with 400 nodes, the largest among the reviewed works.

For SDVRP, Ulmer et al. (2020) argue that it is a more convenient model, which also aligns better with popular approaches to this problem, that the action contains also the route plan information. They define a new variant of MDP, called route-based MDP, in which the state includes the route plan from the last epoch, and the action contains the updated route plan in addition to the next stop to visit. The 'immediate' reward becomes the difference in route value between the old and new plans. Following this line, Joe and Lau (2020) model a system agent whose state includes the cost for the remaining route for each vehicle, and the agent assigns a new request to a vehicle at each decision epoch. The rerouting after matching is solved by simulated annealing for VRP. Under this framework, one only needs to learn an action-value function to generate the matching decisions. The algorithm is tested on a multi-vehicle SDVRP with pick-up/delivery and time windows.[1] In a somewhat similar spirit but for static CVRP, the MDP action in Delarue et al. (2020) is to generate one route (tour). The value network consists of dense layers and ReLU activation and is representable by mixed-integer linear constraints so that the action can be computed through solving a Prize Collecting TSP by MIP.

7.2.4   Connection with Ridesharing

The connection between VRP and ridesharing exists at both local and fundamental levels. As a subproblem in ride-pooling, the rerouting problem after a new passenger is matched to the vehicle is a TSP with pick-up and delivery (TSPPD), which is one-vehicle single-tour instance of CVRP with pick-up and delivery. At a fundamental level, (multi-vehicle) ride-pooling is a stochastic dynamic multi-vehicle CVRP with pick-up and delivery. Although there are no explicit time windows, cancellation may occur if waiting time is too long. Ride-hailing is also a special case where the vehicles all have unit capacity, and in this case, matching and routing merge into one single problem. So the ridesharing problem is an SDCVRP, except that repositioning is an intervention strategy not considered in SDCVRP.

---

[1] This method can be regarded as one for the matching problem in ride-pooling described in Chap. 8.

Ride-Pooling (Carpool)

Ride-pooling optimization typically concerns with matching, repositioning, routing (see e.g., Zheng et al. 2018; Alonso-Mora et al. 2017a, 2017b; Tong et al. 2018). The RL literature has primarily focused on the first two problems. The ride-pooling matching problem differs from that in Chap. 5 in that a combination of multiple passengers, and hence their combined trip, can be matched to a vehicle that may or may not be empty. See stages B and C in Fig. 8.1 from Alonso-Mora et al. (2017a) for an illustration. The repositioning problem is similar to the ride-hailing case, except that the objective is to optimize some pooling-specific metrics that we define next. The routing problem solves for the sequence of pick-ups and drop-offs given the assigned passengers for a vehicle. The routing problem could also concern with route guidance on the road network. See stage D in Fig. 8.1.

8.1 Metrics

Many works have multiple objectives and define the reward as a weighted combination of several quantities, with hand-tuned weight parameters. *Passenger wait time* is the duration between the request time and the pick-up time. *Detour delay* is the extra time a passenger spends on the vehicle due to the participation in the ride-pooling. In some cases, these two quantities define the feasibility of a potential pooled trip instead of appearing in the reward (Shah et al. 2020). *Effective trip distance* is the travel distance between the origin and destination of a trip request, should it be fulfilled without ride-pooling. Yu and Shen (2019) consider minimizing passenger wait time, detour delay, and lost demand. Guériau and Dusparic (2018) maximize the number of passengers served. Jindal et al. (2018) maximize the total effective trip distance within an episode, which is just the number of served requests weighted by individual trip distance. Considering a fixed number of requests within an episode (hence fixed maximum effective distance), this metric reflects the efficiency of

Z. (Tony). Qin et al., *Reinforcement Learning in the Ridesharing Marketplace*, Synthesis Lectures on Learning, Networks, and Algorithms, https://doi.org/10.1007/978-3-031-59640-7_8

ride-pooling. Alabbasi et al. (2019), Haliem et al. (2020, 2021), Singh et al. (2021) all attempt to minimize the SD mismatch, passenger wait time, reposition time, detour delay, and the number of vehicles used. In addition, Haliem et al. (2020) consider the fleet profit, and Singh et al. (2021) study a more general form of ride-pooling, where a passenger can hop among different vehicles to complete a trip, with each vehicle completing one leg. They further consider the number of hops and the delay due to hopping.

8.2   Modeling

The state of an agent usually consists of global SD information, similar to that for matching and reposition, but the vehicle status contains key additional information of occupancy and OD's of the passengers on board.

The action space depends on whether the agent is modeled at vehicle level or system level. Existing RL-based works all require that a vehicle drops off all the passengers on board according to a planned route before a new round of pooling. An individual vehicle agent can then match to a feasible group of passengers (in terms of capacity and detour delay) (Jindal et al. 2018), reposition to another location (Alabbasi et al. 2019; Haliem et al. 2020, 2021), or both (Guériau and Dusparic 2018). A system-level agent has to make action decisions for the entire fleet together (Yu and Shen 2019; Shah et al. 2020). The feasible combinations of passengers are typically determined by a separate process based on a pairwise shareability graph or a trip-vehicle graph (Alonso-Mora et al. 2017a). (See illustration in Fig. 8.1.)

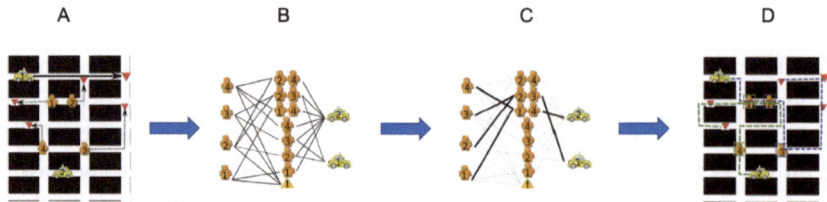

Fig. 8.1 Illustration of the ride-pooling matching process adapted from Alonso-Mora et al. (2017a). Stage A shows the state of the current vehicles and requests. Vehicle 1 has one passenger on board with her destination at the top right-hand corner, while vehicle 2 is empty. The feasible combinations of passengers and the vehicles feasible to serve them are determined at stage B. The corresponding assignment graph is set up and solved at stage C. Stage D shows the resulting routes to fulfill all four new requests as well as the existing trip: Vehicle 1 serves passengers 2 and 3 in addition to the passenger already onboard. Vehicle 2 serves passengers 1 and 4

8.3 Approaches

Papers with vehicle-level policy commonly train a single agent and apply to all the vehicles independently (see e.g., Haliem et al. 2021). DQN is a convenient choice of training algorithm for this setting. For system-level decision-making, both (Yu and Shen 2019 and Shah et al. 2020) employ an ADP approach and consider matching decisions only. Yu and Shen (2019) follow a similar strategy as Simao et al. (2009) and use a linear approximation for the value function. In contrast, Shah et al. (2020) decompose the system value function into vehicle-level ones and adopts a neural network for the individual value function, which is updated by mini-batch stochastic gradient descent similar to that in DQN.

It has become increasingly clear that dynamic routing and route planning in the context of ride-pooling require specific attention. In particular, there are two aspects unique to ride-pooling. First, the trips are known only at their request times. Hence, the routes taken by the pooled vehicles (i.e., the sequences of pick-ups and drop-offs) have to be updated dynamically to account for the newly joined passengers. Tong et al. (2018), Xu et al. (2020) formulate the route planning problem for ride-pooling and develop efficient DP-based route insertion algorithms for carpool. In Sect. 9, we will see that this is also part of the stochastic dynamic vehicle routing problem. Second, within a given route plan, the route taken by a pooled vehicle from an origin to a destination can affect the chance and quality of its future pooling. Hence, dynamic routing (or route choice) between an OD pair can be optimized in that direction, e.g., Yuen et al. (2019) go beyond the shortest-path to make route recommendations for better chance of pooling. Guériau et al. (2020) evaluate the SAMoD system proposed in Guériau and Dusparic (2018) in a microscopic environment based on SUMO with a traffic congestion-aware (non-RL) routing component. We expect to see more RL-based algorithms for the ride-pooling dynamic routing problems.

8.4 Case Study: Single-Driver Ride-Pooling Decision-Making

We present a case study based on the work of Jindal et al. (2018), demonstrating the optimization of high-level ride-pooling decisions using reinforcement learning (RL), with a specific focus on the Deep Q-Learning Network (DQN).

8.4.1 Problem Formulation

We formulate the ride-pooling problem from the perspective of a driver through an MDP.

State A state, denoted $s_i := (l_i, t_i)$, represents the i-th state of an agent (taxi/driver). Here, l_i is a 2-D tuple representing the GPS coordinates (Lat_i, Lon_i), and t_i specifies the time of the day in seconds.

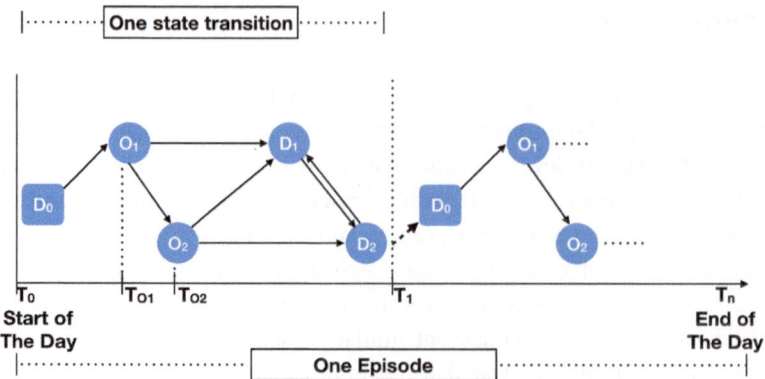

Fig. 8.2 State Transition Process (Jindal et al. 2018)

Action An action, $a \in (W, TK1, TK2)$ has three possible values: W indicates the *wait* action, $TK1$ is the action of assigning a non-carpooling, single passenger termed as *take one* action, and $TK2$ corresponds to *carpool*. We assume that a taxi can be assigned at most two taxi calls. Moreover, these actions correspond to the top-level carpool decisions, and we leave the low-level trip assignments to the environmental in this work.

Reward We define the reward, r, as the effective distance traveled by the taxi throughout a transition. We calculate the effective distance of any trip as the sum of actual distances from the origin of the trips to the individual trip destinations, based on historical data.

Episode One complete day, from 0:00 AM to 23:59 PM, is considered as one episode. An episode completes when the t component of the state of the taxi reaches 23:59 PM.

State Transition The state of the taxi gets updated when it completes an assigned action, and we refer to this change as the *state transition*. As illustrated in Fig. 8.2, one episode is defined by the transition from the start of the day (T_0) to the end of the day or end of the episode (T_n), with $T_0 \rightarrow T_1$ being the first state transition.

8.4.2   Travel Time Estimation

Travel time is the duration taken by a vehicle to move from one location to another, while travel distance is the distance traversed between the two locations. We define a taxi trip $\mathbf{p}_i$ as a 5-tuple $(\mathbf{O}_i, \mathbf{D}_i, t_i, d_i, T_i)$ where the trip starts at the origin $\mathbf{O}_i$ at a given time-of-day t_i, headed to destination $\mathbf{D}_i$, with d_i being the travel distance, and T_i being the travel time. Both the origin and destination are 2-tuple GPS coordinates, and time-of-day t_i is in seconds. An input to our system, defined as a query $\mathbf{q}_i$, is a given pair of (*origin, destination, time-of-day*)$_i$, and the output is the corresponding pair *(travel time, travel distance)*$_i$. Given the historical database of N taxi trips $\mathcal{X} = \{\mathbf{p}_i\}_{i=1}^N$, our goal is to estimate the travel distance and time, (d_q, T_q), for a query $\mathbf{q} = (\mathbf{O}_q, \mathbf{D}_q, t_q)$.

8.4.3   Spatiotemporal Network

Estimation of travel time between two points is a critical factor for accurate simulation in the carpool training environment. Here, we introduce our approach based on deep neural networks for learning travel time for origin-destination pairs not included in the NYC dataset.

The proposed Spatiotemporal Network (ST-NN) architecture, shown in Fig. 8.3, includes two distinct deep neural network (DNN) modules, namely the "Dist-DNN Module" for travel distance estimation and the "Time-DNN Module" for travel time estimation. The Dist-DNN module receives the binned GPS coordinates of the origin and destination as input, while the Time-DNN function takes the activations of the last hidden layer of the Dist-DNN module (the encoded feature vector) and the time-of-day information as input.

The ST-NN architecture is then trained via stochastic gradient descent jointly for both travel distance and time according to the loss function:

$$L(Y_D, Y_T, \hat{Y}_D, \hat{Y}_T) = \frac{1}{2N} \sum_{i=1}^N (Y_T^i - \hat{Y}_T^i)^2 + \frac{1}{2N} \sum_{i=1}^N (Y_D^i - \hat{Y}_D^i)^2. \qquad (8.1)$$

8.4.4   Q-Learning for Ride-Pooling Decisions

Given that the taxi is entirely reliant on RL for its carpooling decision-making, learning the value function of the taxi's state-action pair from experience gathered from the carpooling simulator is crucial. We adopt a model-free RL approach to learn an optimal policy since

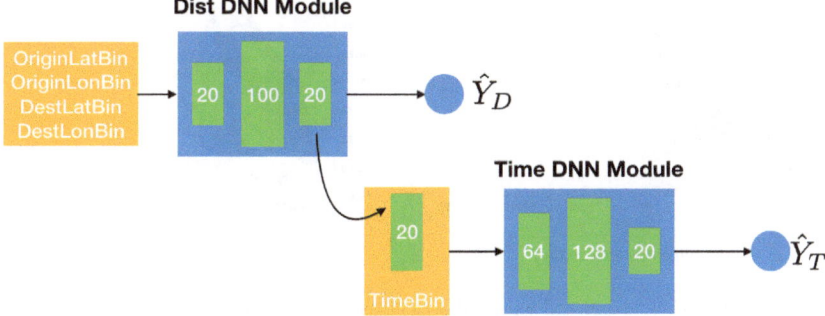

Fig. 8.3 Unified neural network architecture for joint estimation of travel time and distance (Jindal et al. 2018)

the agent has no knowledge of state transitions and reward distributions. We employ a function approximator to model $Q(\mathbf{s}, a) = f_\theta(\mathbf{s}, a)$ in dealing with large state-action space or continuous state space scenarios. This is done using a three-layer deep neural network that learns the state-action value function, as proposed in Mnih et al. (2015). To ensure training stability, we use Double-DQN (Van Hasselt et al. 2016), where a target Q-network $\hat{Q}$ is maintained and synchronized periodically with the original Q-network.

8.4.5   Simulation Results

Applying the ST-NN, we incorporated the trip time estimation module into our carpool training environment. We then trained an RL agent using this simulated environment to optimize the carpooling policy of a single taxi driver. We evaluate the performance of the DQN learned policy by comparing the mean cumulative reward with respect to a fixed policy that always favors carpooling and the tabular-Q policy. The DQN learned policy outperformed both the fixed policy and the tabular Q policy for both weekday and weekend scenarios.

Two different regions in New York City (NYC) were selected for this study: Uptown Manhattan (with a lower taxi call density) and Downtown Manhattan (with a higher taxi call density) as shown in Fig. 8.4. In both the selected regions, the DQN policy outperforms the fixed policy and Tabular Q, indicating that the DQN has successfully learned an optimal policy for ride-pooling decisions.

In Downtown Manhattan, where taxi calls are very frequent, the DQN policy always favors carpooling and generates rewards similar to those of the fixed policy. However, in Uptown Manhattan, where taxi calls are less frequent, the DQN learned policy selectively takes $TK1$ or W action, leading the taxi into regions with higher long-term values. This behavior

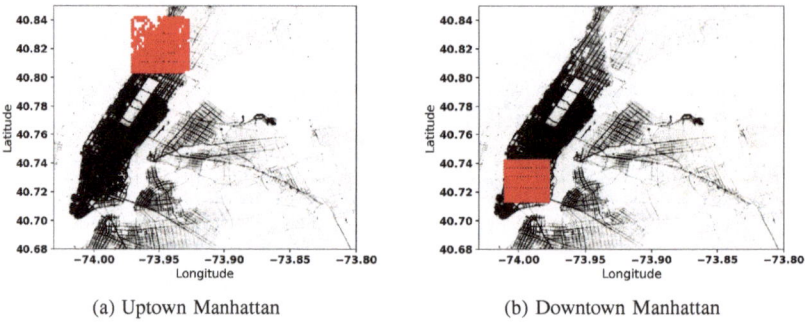

(a) Uptown Manhattan   (b) Downtown Manhattan

Fig. 8.4 Selected regions visualized with red dots (Best seen in color) (Jindal et al. 2018)

Table 8.1 Mean cumulative reward on weekday and weekend

Region	Day	Fixed policy	Tabular Q	DQN
Uptown	Weekday	41.543	39.17	**46.08**
	Weekend	25.39	14.37	**27.86**
Downtown	Weekday	**340.06**	186.00	339.42
	Weekend	259.57	145.63	**261.23**

indicates that our proposed RL technique for ride-pooling not only learns to maximize the current reward, but also strategically plans to position the vehicle for future rewards.

Key simulation results are summarized in Table 8.1.

Related Methods

9

9.1 Causality and Reinforcement Learning

There has been increasing interest in investigating the interplay of causality and RL in the recent years. The synergy of the two becomes understandable by treating the task of learning the action values $Q(s, a)$ in RL as estimating the long-term counter-factual effects of applying the different actions to a given current state, to which knowing the corresponding causal structure in the environment is highly helpful. An example of this perspective is model-based RL as a causal inference problem (Gasse et al. 2021). In Chap. 3, we have introduced the concept of *causal RL*, a major direction of combining causality with RL.

In this section, we introduce representative works leveraging causality and RL for ridesharing via using RL for estimating causal effects in AB experiments and using causal modeling for off-policy evaluation (OPE) in the context of a two-sided marketplace.

9.1.1 Experimentation

For A/B testing marketplace-level ridesharing system policies, common user-split tests suffer from serious interference due to the sharing of the same rider or driver pool and the interaction between the riders and drivers. Consider testing a new dispatching algorithm. It does not work to simply divide the riders or drivers into the control and treatment groups and applying the two different policies because they share the other side of the marketplace, which invalidates the stable unit treatment value assumption (SUTVA). Hence, practitioners typically resort to switch-back tests (also known as time-split tests) (Bojinov et al. 2023) to estimate the treatment effects of system policies. Specifically, under this kind of experimentation design, only one algorithm is run in the marketplace at any time, and the two candidate algorithms are usually run alternatingly in a sequential manner over the experiment period.

© The Author(s), under exclusive license to Springer Nature Switzerland AG 2025
Z. (Tony). Qin et al., *Reinforcement Learning in the Ridesharing Marketplace*,
Synthesis Lectures on Learning, Networks, and Algorithms,
https://doi.org/10.1007/978-3-031-59640-7_9

This avoids the cross-group interference arising from contention that we explained above. However, rideshare system policies often have long-term effects that extend beyond the policy switch boundaries, resulting in *spill-over*. In addition, many algorithms, such as RL, depend on past system states to generate the next decisions (actions), leading to *Markovian interference*. Hence, estimating the treatment effects of a ridesharing switch-back experiment is a challenging task.

Recently, there has been an interesting line of works that treat the experiment itself as a sequential decision problem. Specifically, the (switch-back) experiment is modeled by an MDP, where state $s_t \in S$ is the system state at time t, and the action $a_t \in \mathcal{A}$ at each decision point is whether to apply the treatment algorithm (intervention) or not. The goal is to estimate the average treatment effect (ATE), which is the difference between the total (or average) returns of the control and treatment policies over the experiment duration should only one of them is executed throughout. The ATE is defined by

$$ATE := \lambda^{\pi_1} - \lambda^{\pi_0}, \tag{9.1}$$

where λ^{π} is the returns of the policy π.

The method proposed in Farias et al. (2022) is an on-policy method that compares the average rewards between the control and treatment policies, i.e., $\lambda^{\pi} = \lim_{T \to \infty} \frac{1}{T} \sum_{t=1}^{T} r(s_t, \pi(s_t))$. With the Q-function as the state-action value function of the MDP, Farias et al. (2022) propose the Difference-in-Q's (DQ) estimator,

$$\hat{ATE}_{DQ} = \frac{1}{T_1} \sum_{t \in T_1} \hat{Q}_{\pi_{1/2}}(s_t, a_t) - \frac{1}{T_0} \sum_{t \in T_0} \hat{Q}_{\pi_{1/2}}(s_t, a_t), \tag{9.2}$$

where $\hat{Q}_{\pi_{1/2}}$ is an estimator of the Q-function under the experiment (switch-back) policy $\pi_{1/2}$ learned from regression.

On the other hand, Shi et al. (2023b) define the value functions in the counterfactual sense. For any $t \geq 0$, let $\bar{a}_t = (a_0, a_1, \ldots, a_t)^\top \in \{0, 1\}^{t+1}$ denote a treatment history vector up to time t. And, for any $(\bar{a}_{t-1}, \bar{a}_t)$, let $S_t^*(\bar{a}_{t-1})$ and $Y_t^*(\bar{a}_t)$ be the counterfactual state and counterfactual outcome, respectively, that would occur at time t had the agent followed the treatment history $\bar{a}_t$. The goodness of a policy π is measured by the discounted cumulative rewards,

$$\lambda^{\pi} = V(\pi; s) = \sum_{t \geq 0} \gamma^t \mathbb{E}\{Y_t^*(\pi) | S_0 = s\},$$

where $0 < \gamma < 1$ is a discount factor, and $V(\pi; s)$ is defined through potential outcomes rather than the observed data. Then, the ATE to estimate is

$$ATE_{CRL} = \int_s \{V(\pi_1; s) - V(\pi_0; s)\} \mathbb{G}(ds), \tag{9.3}$$

for a given reference distribution function $\mathbb{G}$ that has a bounded density function on $\mathbb{S}$. The ATE estimator is computed based on a version of temporal-difference learning applying basis function approximation.

Both works have shown asymptotic normality properties of their proposed estimators and tested their approaches on ride-hailing data sets. Li et al. (2023) extends the switch-back experiment design to a dynamic treatment assignment and investigates the optimal treatment allocation policy within a similar MDP setup as in the previous works.

9.1.2 Off-Policy Evaluation (OPE)

OPE is a long-standing challenge in RL. An additional complexity in the context of (rider) incentives policy for ridesharing is unmeasured confounders. The major complication is that the target policy value is defined based on the intervention distribution and cannot be easily approximated through the distribution of the observed data. One common modeling choice is to cast the problem into a POMDP OPE problem, existing works on which, however, are unable to quantify the uncertainty in the value estimates. Shi et al. (2022) considers the setting of a confounded infinite horizon MDP with mediators, where the unmeasured confounder $U_t \sim p_u(\bullet|S_t)$, the action is generated by $A_t \sim p_a(\bullet|S_t, U_t)$, and the mediator M_t is generated using $p_m(\bullet|A_t, S_t)$ which is not confounded by U_t. The agent receives a reward $R_t \sim p_r(\bullet|M_t, A_t, S_t, U_t)$ and the environment transits into the next state $S_{t+1} \sim p_s(\bullet|M_t, A_t, S_t, U_t)$. The mediators are intermediate variables that mediate the effect of actions on the system dynamics. In the ride-hailing context, the mediator corresponds to the final discount applied to each ride. Notably, the final discount might be different from the discount included in the incentives program, as it depends on other promotion strategies the platform applies to the ride, but is conditionally independent of other unmeasured variables that confound the action. In addition, the action will affect the immediate reward and future state variables only through the mediator. Shi et al. (2022) show that with the presence of the mediators, the value of the target policy η^π is identifiable in the observational data under the influence of unmeasured confounders. The observational data consist of N i.i.d. trajectories, summarized as $\{(S_{i,t}, A_{i,t}, M_{i,t}, R_{i,t}, S_{i,t+1})\}_{1 \le i \le N, 0 \le t < T_i}$ where T_i corresponds to the termination time of the ith trajectory. The corresponding (state) value function as

$$V^\pi(s) = \sum_{t=0}^{+\infty} \gamma^t \mathbb{E}^\pi(R_t|S_0 = s), \tag{9.4}$$

where the expectation $\mathbb{E}^\pi$ is defined by assuming the system follows the policy π. Based on the observed data, the objective is to learn the aggregated value $\eta^\pi = \mathbb{E}\{V^\pi(S_0)\}$ where the expectation is taken with respect to the initial state distribution, and to construct its associated confidence interval.

There are two existing common estimators that one can employ for η^π. The Direct Estimator estimates the Q-function (state-action value function, Q^π) based on the observed data and uses it to estimate the value of a target policy. Unfortunately, this estimator carries a substantial bias due to potential model misspecification. The Importance Sampling (IS) estimator leverages the stationary property of the state transitions to break the curse of high variance typical in sequential decision making (Liu et al. 2018). However, the IS estimator requires correction specification of the marginal density ratio ω^π and the mediator distribution p_m, and it also suffers from high variance due to the inverse probability weighting.

The proposed estimator combines the two estimator above, addressing the challenges in both. Specifically, it takes the following form,

$$\widehat{\eta} = \frac{1}{\sum_i T_i} \sum_{i=1}^{N} \sum_{t=0}^{T_i-1} \psi(O_{i,t}), \tag{9.5}$$

where $O_{i,t} = (S_{i,t}, A_{i,t}, M_{i,t}, R_{i,t}, S_{i,t+1})$ is the observed transitions data, $\psi(O)$ is the estimating function which is a combination of the direct estimator and three augmentation terms in the estimates of $(Q^\pi, \omega^\pi, p_a^*, p_m)$. A Wald-type confidence interval for the estimated policy value can then be constructed, offering the necessary tools for uncertainty quantification. Shi et al. (2022) additionally show the asymptotic properties of the proposed estimator, including consistency and asymptotic correctness of the confidence interval.

9.2   Approximate Dynamic Programming

A family of methods closely related to RL is approximate dynamic programming (ADP) (Powell 2007) for solving stochastic dynamic programs (DP), of which the Bellman equation for MDP is an instance. In ADP methods, unlike that typically seen in RL, a post-decision state s_t^x is often defined to represent the intermediate state to which the current state s_t will transition deterministically given the action a_t before the random factors ω_t (e.g., demand appearance and cancellation) in the environment realize. With ω_t fully realized, the state transitions into the next pre-decision state s_{t+1}. The value function in an ADP method is defined on the post-decision state and is approximated by a particular functional form. Given the approximated values, the original optimization problem is solved to obtain the decision solution for the current time step. Linear function approximation is popular (e.g., Simao et al. 2009; Yu and Shen 2019; Al-Kanj et al. 2020) because the dual variables associated with the solution to the current-stage optimization can be used to update the linear function parameters. Then, the state is advanced to the next pre-decision state, and the iteration continues until convergence. By nature, ADP methods are on-policy methods. Recently, neural network-based value function approximation (Shah et al. 2020) has also been adopted and developed due to their higher level of flexibility. In this case, the value function updates largely follow the DQN scheme. The ADP methods for ridesharing reviewed in this book solve system-level stochastic DP problems (e.g., matching and repositioning) and aim to approximate the system value by decomposing it into local or driver-centric values, and the update schemes employed fall into the family of approximate value iterations.

9.2.1 RL for Ridesharing Through an Alternative Lens

ADP offers an interesting alternative lens to interpret the underlying mechanism of the RL algorithms for rideshare. Dispatching is the operation that underpins a large part of the rideshare marketplace, and it is also the primary venue where the actual monetary transactions take place. Hence, we use dispatch as the canonical example for illustrating our point. The exposition on the ADP part (the next two subsections) is based on Simao et al. (2009), Al-Kanj et al. (2020).

9.2.1.1 Dispatch Optimization

The basic partitioning of the spatial space is the grid system (geohash or hexagonal cells). The grid cell of a driver is denoted by a, and the origin-destination (OD) of a request (cell-level route) is d, consisting of the OD cells. The decision that the system can impose on each vehicle is also denoted by d for notational simplicity, either to service a trip (dispatch) or idle (assuming to stay at the current cell).

The optimization problem for dispatch aims to maximize the total expected discounted cumulative returns within the optimization horizon subject to supply and demand constraints:

$$\max_{\pi \in \Pi} \mathbb{E}\left[\sum_{t=0}^{T} \gamma^t C(S_t, X_t^{\pi}(S_t)) \Big| S_0\right] \tag{9.6}$$

$$s.t. \sum_{d \in D(a)} x_{tad} = R_{ta}, \ \forall a \in \mathcal{A}$$

$$\sum_{a \in A(d)} x_{tad} \leq D_{td}, \ \forall d \in \mathcal{D}$$

$$x_{tad} \geq 0, \ \forall a, d$$

$X_t^{\pi}(S_t)$ is a function that determines x_t given S_t, i.e., the solver for the assignment problem. Let χ_t be the feasible set at time t (including the supply, demand, and non-negativity constraints).

The original stochastic optimization problem (9.6) is hard to solve over multiple time periods. The Bellman equation for the optimal policy offers a view to solve the problem step-wise,

$$X_t^*(S_t) = \arg\max_{x_t \in \chi_t} C_t(S_t, x_t) + \gamma \mathbb{E}\left[V_{t+1}(S_{t+1}) \mid S_t, x_t\right] \tag{9.7}$$

A system state transition is modeled by two stages. The state St first deterministically transitions into a *post-decision state* S_t^X under the effect of the action x_t, e.g., vehicles transported by rides. We then observe the realization of exogenous information (e.g., demand in the next period), and S_t^X transitions into S_{t+1}. Stochasticity is involved in the second stage. Using S_t^X allows us to get rid of the expectation in the first-stage optimization,

$$X_t^*(S_t) = \arg\max_{x_t \in \chi_t} \; C_t(S_t, x_t) + \gamma V_t^X(S_t^X), \tag{9.8}$$

where $V_t^X(S_t^X) := \mathbb{E}\left[V_{t+1}(S_{t+1}) \mid S_t^X\right]$, and $V_t(S_t) := \max_{x_t \in \chi_t} C_t(S_t, x_t) + \gamma V_t^X(S_t^X)$.

The above suggests a two-step iterative algorithm for solving this multi-period stochastic optimization problem:

1. Solve for the decisions of the current time step, using an approximation of the value function, V_t^X.
2. Update the approximation of the value function, V_t^X.

9.2.1.2   System Value Function Approximation and Update

The value function (expected future returns given the current system state) is approximated by a function linear in the resource:

$$V_t^X(S_t^X) \approx \overline{V}(S_t^X) = \sum_{s'} \overline{v}_{t'(s,d),s'}^n \sum_{s,d} \delta_{s'}(s,d)x_{tsd} = \sum_{s,d} \overline{v}_{t',s'}^n x_{tsd} \tag{9.9}$$

$\delta_{s'}(s,d)$ is an indicator function which is equal to 1 if a driver in state a transitions into s' under action d and 0 otherwise. We have simplified the notation to use t', s', with the understanding that they are the results of executing decision d on state s, i.e., $t' \equiv t'(s,d)$, $s' \equiv s'(s,d)$. t' takes into account pick-up and trip time of fulfilling the ride d, and s' is the destination location of the ride. $\overline{v}_{t',s'}^n$ represents the expected marginal value of a driver in s' at time t' from the n-th iteration. It is a linear approximation of the slope at a point of an otherwise piecewise linear function. In ADP, it is iteratively updated by the dual variables (which is the current realization of the marginal values) corresponding to the resource constraints of the optimization problem as follows:

Each iteration n is an episode where the time runs from $t = 1, \ldots, T$. The step-t optimization problem in the n-th episode is then

$$\max_{x_t \in \chi_t} \sum_{s,d} \left(c_{tsd} + \gamma \overline{v}_{t',s'}^{n-1}\right) x_{tsd}. \tag{9.10}$$

That is, to solve the current cycle assignment problem using the current approximation of the supply values $\overline{v}$, where $\sum_{s,d} \overline{v}_{t',s'}^n x_{tsd}$ forms a first-order decomposition of the system value at the post-decision state. This is essentially the matching problem solved in production in each cycle albeit at a coarser (grid cell) level. In the forward-pass algorithm of ADP for the dispatch problem, the dual variables $u_{t,s}$ of this LP are used to update $\bar{v}_{t,s}^{n-1}$ in the approximate value iteration,

$$\bar{v}_{t,s}^{n} \leftarrow (1 - \alpha_n)\bar{v}_{t,s}^{n-1} + \alpha_n u_{t,s}, \tag{9.11}$$

where n is the learning rate at n-th iteration. It is then used in the next iteration to solve for the assignment decisions. We can maintain an approximation to the marginal demand values in a similar way by using the dual variables u_{td}, which are only by-products and do not influence the future decisions in this framework.

9.2.1.3 A View of System Value Decomposition

Let u_s and u_d be the dual variables associated with the supply and demand constraints, respectively. The dual values represent the cumulative marginal values over a future horizon up to T. Then, the dual problem to the cycle-t assignment problem (9.10) is

$$\min_{u_{ts}, u_{td}} \sum_{s} R_{ts} u_{ts} + \sum_{d} D_{td} u_{td} \tag{9.12}$$

$$s.t. \ u_{ts} \text{ free}, u_{td} \geq 0$$

$$u_{ts} + u_{td} \geq c_{tsd} + \gamma \bar{v}_{t',s'}^{n-1}, \ \forall d, s \mid (s, d) \in \delta(r),$$

where $\delta(r)$ is the neighborhood area defined by the dispatch radius.

By LP duality, once (9.10) is solved (and so is (9.12)), the objective value of (9.12) is equal to that of (9.10), which is the system value $V_t(S_t)$, i.e.,

$$V_t(S_t) := \max_{x_t \in \chi_t} \sum_{s,d} \left(c_{tsd} + \gamma \bar{v}_{t',s'}^{n-1} \right) x_{tsd} = \sum_{s} R_{ts} u_{ts}^* + \sum_{d} D_{td} u_{td}^*. \tag{9.13}$$

This presents a convenient view of allocation of the marginal contributions (cumulative over T) from the supply and demand units at time t, which are given by (u_{ts}^*, u_{td}^*). Furthermore, by the complementary slackness conditions of LP, we have for the matching options (s, d) selected by the solver,

$$u_{ts}^* + u_{td}^* = c_{tsd} + \gamma \bar{v}_{t',s'}^{n-1}. \tag{9.14}$$

In the steady-state, $\bar{v}_{t',s'}^{n-1} = \bar{v}_{t',s'}$, and $\mathbb{E}[u_{ts}^*] = \bar{v}_{ts}$. Hence, the cumulative marginal demand value $u_{td}^* \approx c_{tsd} + \gamma \bar{v}_{t',s'} - \bar{v}_{ts}$, which is the temporal difference error or advantage associated with the match.

9.2.1.4 Connection to RL

The dispatch problem is a system-level problem, and hence, it is natural to think about system-level modeling, with system state S_t and system action (assignment) A_t. However, this view presents a mounting challenge of combinatorial actions. Therefore, a more common

approach to this problem is through a driver-oriented view by defining an MDP for a driver
as we have described in Sects. 5.8 and 5.9. Correspondingly, we have $V(s)$ for the driver
(supply) state value function to be shared by all drivers and $Q(s, d)$ for the driver action-value
function. In each dispatch cycle, we optimize

$$\max_{\{s_t, d_t\} \in \chi_t} \sum_{(s_t, d_t)} Q(s_t, d_t) \approx \sum_{s,d} \left(c_{tsd} + \gamma V(s'_{t'}) \right) x_{tsd}, \tag{9.15}$$

where $c_{tsd} + \gamma V(s'_{t'})$ is a sample approximation to $Q(s_t, d_t)$, and s' is the next-state after
executing action d.

Since the dispatch system is a multi-agent system where the agents (drivers) interact
with each other, the independently learned supply value functions (through realized trip
experiences) do not sum up to the system value in general. In fact, we have shown that the
total OSV summed over all open supplies in a cycle is generally an overestimate of the
system value.

What we desire is to learn the appropriate unit value functions $Q(s_t, d_t)$ and $V(s_t)$ such
that

$$Q(S_t, A_t) = \sum_{(s_t, d_t) \in A_t} Q(s_t, d_t) = \sum_{(s_t, d_t) \in A_t} V(s_t) + A(s_t, d_t), \tag{9.16}$$

where $A(s, d) := Q(s, d) - V(s)$ is the advantage of action d at state s. It measures the
incremental value of taking a particular action (i.e., matching to request d) as compared to
the average case. $(V(s) = \mathbb{E}_{d \sim \pi(s)} Q(s, d).)$

From the previous two sections, we see that if we use a supply value function iteratively
updated by the duals (i.e., the cumulative marginal supply values $\bar{v}$), then we do have the
right decomposition of the system value in the steady state. Furthermore, since all open
supplies in the cycle are matched (some possibly matched to idling),

$$\max_{A_t} \sum_{(s_t, d_t), d_t \in A_t} V(s_t) + A(s_t, d_t) \equiv \max_{A_t} \sum_{(s_t, d_t), d_t \in A_t} A(s_t, d_t), \tag{9.17}$$

since $V(s_t)$ is independent from the action decisions. We see that what we are effectively
doing is maximizing the total advantage of the assignments, in other words, approximately
maximizing the total cumulative marginal demand values in these assignments. Intuitively,
it is a process of finding a set of assignments with the cycle such that by doing so we gain the
maximal amount of extra future value as compared to the average case. The total incremental
value here is the cumulative marginal demand values.

9.3   Model Predictive Control

A related family of forward-looking methods that share a similar spirit with model-based RL
is model predictive control (MPC). MPC exploits the environment dynamics model more
explicitly in that it solves a multi-step planning problem over the prediction horizon at deci-

sion time and executes the first-step decision. The process is repeated in a rolling-horizon fashion. For ridesharing applications, both MPC and model-based RL involve online prediction of supply and demand using models trained on historical data. Model-based RL uses the prediction to learn the environment model, whereas MPC utilizes the information to generate the planning problem, which is typically solved by a mixed-integer-linear-programming (MILP) solver.

MPC-based repositioning methods have been developed for both regular ridesharing (Iglesias et al. 2018; Zhang et al. 2016; Miao et al. 2016; Cheng et al. 2018) and ride-pooling (Riley et al. 2020; Miller and How 2017). These works solve an MILP for online planning and separately-trained SD forecast models. Zhang et al. (2016), Iglesias et al. (2018), Cheng et al. (2018) solve for zone-level decisions. Zhang et al. (2016) additionally considers charging constraints for electric vehicles. Miao et al. (2016), Miller and How (2017) directly solves for vehicle-level decisions, and Riley et al. (2020) solves two optimization problems, the second of which assigns individual vehicles according to the plan generated by the first.

9.3.1   Incorporating Reinforcement Learning

A recent work (Wei et al. 2023) explicitly incorporates RL elements into an MPC-type framework. The reposition planning model adopts a fluid-based formulation (Braverman et al. 2019), which generates the look-ahead reposition policy through a linear program that accounts for driver non-compliance, which is a key practical challenge in driver repositioning. The spatiotemporal driver value network from Sect. 6.2.3 is incorporated into the LP objective function to broaden the planning horizon beyond the time slots explicitly modeled in the LP.

9.3.1 Incorporating Reinforcement Learning

Open Resources

10

10.1 Data Sets

The problems in ridesharing are highly practice-oriented, and results from toy data sets or environments may present a very different picture from those in reality. Hence, real-world data sets and realistic simulators backed up by them are instrumental to research in RL algorithms for these problems.

The most commonly used data sets are those made available by NYC TLC (Taxi & Limousine Commission) (TLC 2020). This large public data repository contains trip records from several different services, Yellow Taxi, Green Taxi, and FHV (For-Hire Vehicle), from 2009 to 2020. The Yellow Taxi data is the most frequently used for various studies. The FHV trip records are submissions from the TLC-licensed bases (e.g., Uber, Lyft) and have a flag indicating pooled trips offered by UberPool and Lyft Shared. The pick-up and drop-off locations are represented by taxi zones. Manhattan, for example, is divided into 64 zones. There is no driver ID associated with the trip records, so reconstructing historical driver-based trajectories is not possible. An older version of the NYC data set (Donovan and Work 2016), however, does include GPS coordinates for pick-up and drop-off locations, and car IDs can be used to track drivers within each year, allowing for more granular and diverse analyses. A similar subset of the NYC FHV data is also available at Kaggle (2017), with GPS coordinates for pick-up and drop-off locations. In addition, travel time data between OD pairs can be obtained through Uber Movement (Uber 2021).

Another taxi data set is the San Francisco data set Piorkowski et al. (2009), which contains GPS coordinates of approximately 500 taxis collected over 30 d in the San Francisco Bay Area in May 2008. The average time interval between two consecutive location updates is less than 10 s.

© The Author(s), under exclusive license to Springer Nature Switzerland AG 2025   89
Z. (Tony). Qin et al., *Reinforcement Learning in the Ridesharing Marketplace*,
Synthesis Lectures on Learning, Networks, and Algorithms,
https://doi.org/10.1007/978-3-031-59640-7_10

A more recent rideshare (Transportation Network Providers, TNPs) data set is published by Chicago Data Portal (2020). This data set contains trips, drivers, and vehicles data reported by Transportation Network Providers (TNP, or rideshare companies) in Chicago from 2018. Trip origin and destinations are represented by census tracts. Times are rounded to the nearest 15 min. Fares are rounded to the nearest $2.50 and tips are rounded to the nearest $1.00. The driver and vehicle data are not joinable with the trip data.

10.2   Simulators

Developing ridesharing simulators has been a line of research itself. Yao and Bekhor (2021) offer a comprehensive review of recent works on ridesharing simulation models, most of them covering a subset of considerations on the number of passengers, the pre-/post-match passenger cancellation behaviors, and driver acceptance/rejection behaviors. In Yao and Bekhor (2021), a sophisticated event-based simulation framework is proposed to capture all aspects of the behavior modeling. Although the 'ridesharing' in their paper is known as the hitch service, where the driver is on her own trip as well, the modeling framework is general and accommodates the ridesharing setting in this survey. Chaudhari et al. (2020a) offer a Gym[1]-compatible, open-source ride-hailing environment (Chaudhari et al. 2020b) for training dispatching and repositioning agents. For large-scale simulation on transport networks, AMoDeus (Ruch et al. 2018) and MATSim (Axhausen et al. 2016) are well-established Java-based simulation frameworks that also come with graphical user interfaces and visualization tools. They are of more sophisticated engineering architectures albeit with higher programming bars for extension. The evaluation simulation environment for the KDD Cup 2020 competition is available for public access through the DiDi decision intelligence simulation platform (DiDi 2021). Although not yet open-sourced, this simulation environment supports both matching and vehicle repositioning tasks and accepts input algorithms through a Python API.

---

[1] https://www.gymlibrary.dev/.

Challenges and Opportunities 11

Given the state of the current literature, we discuss a few challenges and opportunities that we feel crucial in advancing RL for ridesharing.

11.1 Ride-Pooling

As seen in Chap. 8, the reward function in ride-pooling is often a hand-tuned combination of multiple objectives. It is desirable to have a principled way to determine the best weighting scheme automatically, potentially leveraging inverse RL and multi-objective learning techniques (Zou et al. 2021; Arora and Doshi 2021) in a similar sense of the ride-hailing case (Zhou et al. 2021a). Methods for learning to make matching decisions are still computationally intensive (Shah et al. 2020; Yu and Shen 2019), in part due to the need to use VRP solver to determine feasible actions (combination of passengers). Moreover, all existing works assume that the action set is pre-determined, and some make only high-level decisions of reposition and serving new passengers or not. A more sophisticated agent may be called for to figure out, for example, how to dynamically determine the desirable passenger combination to match to a vehicle and the routes to take thereafter. Ride-pooling pricing (Ke et al. 2020a), a hard pricing problem itself, is tightly coupled with matching. A joint pricing-matching algorithm for ride-pooling is therefore highly pertinent. As mentioned in Chap. 8, it is also highly anticipated to go beyond using generic routing algorithms and to tailor them to ride-pooling with RL.

Z. (Tony). Qin et al., *Reinforcement Learning in the Ridesharing Marketplace*,
Synthesis Lectures on Learning, Networks, and Algorithms,
https://doi.org/10.1007/978-3-031-59640-7_11

11.2 Joint Optimization

The rideshare platform is an integrated system, so joint optimization of multiple decision modules leads to better solutions that otherwise unable to realize under separate optimizations, ensuring that different decisions work towards the same goal. RL for joint optimization across multiple modules calls for research on reward function design, state-action representation that facilitates inter-module communication, and the training algorithms. Models and algorithms that allow decentralized execution by the different modules are highly preferred in practice. Joint learning and optimization via a shared value function (Tang et al. 2021), for example, is practically appealing because of its deployment-friendly architecture.

We have already seen development on RL for joint matching-reposition (Holler et al. 2019; Jin et al. 2019; Tang et al. 2021) and with ride-pooling (Guériau and Dusparic 2018), pricing-matching (Chen et al. 2019a), and pricing-reposition (Turan et al. 2020). An RL-based method for fully joint optimization of all major modules is highly expected. Meanwhile, this also requires readiness from the rideshare platforms in terms of system architecture and organizational structure.

11.3 Heterogeneous Fleet

With the wide adoption of electric vehicles and the emergence of autonomous vehicles, we are facing an increasingly heterogeneous fleet on rideshare platforms. Electric vehicles have limited operational range per their battery capacities. They have to be routed to a charging station when the battery level is low (but sufficiently high to be able to travel to the station). Autonomous vehicles may run within a predefined service geo-fence due to their limited ability (compared to human drivers) to handle complex road situations. For an RL-based approach, a heterogeneous fleet means multiple types of agents with different state and action spaces. The adoption of autonomous vehicles also opens new operational paradigms. Dynamic fleet size inflation (Beirigo et al. 2022), for example, hires idle autonomous vehicles on demand to guarantee service quality contracts in a ridesharing marketplace. Specific studies are required to investigate how to make such a heterogeneous fleet cooperate well to complement each other and maximize the advantage of each type of vehicles to improve overall system efficiency.

11.4 Simulation and Sim2Real

Simulation environments are fundamental infrastructure for successful development of RL methods. Despite those introduced in Chap. 10, simulation continues to be a significant engineering and research challenge. We have rarely seen comparable simulation granularity as that of the environments for traffic management, (e.g., SUMO Lopez et al. 2018; Flow Wu et al. 2017) or autonomous driving (e.g., SMARTS Zhou et al. 2020b; CARLA Dosovitskiy et al. 2017).[1] The opportunity is an agent-based microscopic simulation environment for ridesharing that accounts for both ride-hailing and carpool, as well as driver and passenger behavior details, e.g., price sensitivity, cancellation behavior, driver entrance/exit behavior. None of the existing public/open-source simulators supports pricing decisions. Those simulators described in the pricing papers all have strong assumptions on passenger and driver price elasticities. A better way might be to learn those behaviors from data through, e.g., generative adversarial imitation learning (Shang et al. 2019) or inverse RL (Mazumdar et al. 2017).

No publicly known ridesharing simulation environment has sufficiently high fidelity to the real world to allow an agent trained entirely in it to deploy directly to production. Several deployed works (Qin et al. 2020a; Jiao et al. 2021) discussed in this book have all adopted offline RL for learning the state value functions and online planning. The robotics community has been extensively investigating ways to close the reality gap (Traoré et al. 2019; Mehta et al. 2020). Sim2real transfer algorithms for ridesharing agents (see Chen et al. 2021b; Chen et al. 2023 for studies along this direction on driver incentives optimization) are urgently sought after.

11.5 Human Behavior

Central to ridesharing platforms are human participants (passengers and drivers).[2] The impact of human behavior is pervasive in the ridesharing marketplace, e.g., in request conversion, cancellation, idle driver diffusion, driver sign-in and sign-off, rider and driver responses to incentives. Human behavior is inherently stochastic and difficult to model, especially with limited data (in size and features), which introduces errors to optimization and simulation. Compared to traditional approaches from operations research, RL offers potential to better handle these stochasticity issues through its adaptability and data-driven nature.

Unlike cumulative effects induced by spatiotemporal transitions (e.g., matching), human-induced long-term effects from changes in habituation and sentiment on the marketplace are much harder to learn due to the much longer horizon such effects span over. To RL, this is dual challenge and opportunity. The challenge is the long feedback loop and very delayed reward signals, and the opportunities lie in engineering and capturing more refined system state features that capture human behavior characterization better and in designing a richer set of reward signals that facilitate the learning of policies for long-term optimality.

---

[1] Guériau et al. (2020) evaluate the ridesharing algorithms in SUMO, but the environment is not public.

[2] A partial exception is an autonomous ridesharing platform, where the supply side is powered by autonomous vehicles. However, such services are still prototypical at the time of writing and have very limited coverage.

11.6   Non-stationarity

We have seen in Chaps. 5 and 6 that RL algorithms deployed to real-world systems generally adopt offline training—once the value function or the policy is deployed, it is not updated until the next deployment. Value functions trained offline using a large amount of historical data are only able to capture recurring patterns resulted from day-on-day SD changes. However, the SD dynamics can be highly non-stationary in that one-time abrupt changes can easily occur due to various events and incidents, e.g., concerts, matches, and even road blocks by traffic accidents. To fully unleash the power of RL, practical mechanisms for real-time on-policy updates of the value function (e.g., Tang et al. 2021; Tong et al. 2021; Eshkevari et al. 2022) is required. In view of the low risk tolerance of production systems in general, sample complexity, computational complexity, and robustness are the key challenges that such methods have to address.

11.7   Marketplace Rewards

Total trip fares (or total revenue, GMV) are not market volume agnostic. It is hard to compare in GMV between policies run in two different time periods because both the supply and demand volumes may change organically over time. It is important for a reward signal to consider the situations on both sides of the marketplace to truthfully reflect the effectiveness and efficiency of a lever policy. A simple example is that a policy that simply brings in more drivers may improve the service level at the expense of a lower driver utilization rate; similarly, another policy that straightforwardly acquires more riders may improve utilization, but the service may take a hit if the supply is insufficient. Hence, practitioners typically look at a basket of marketplace metrics focused on the rider, driver, and system perspectives. However, it is difficult to synthesize a multi-dimensional reward signal to easily make conclusion on what exactly has happened to the marketplace due to the application of a specific policy. One approach to tackle the challenge is to focus on market balance. Chin and Qin (2023) extends the graph-based equilibrium metrics (GEM) (Zhou et al. 2021b) to a rider-driver dual-perspective framework, within which the trade-off between the rider and driver-side metrics can be better captured and synthesized. A number of issues still remain, e.g., accounting for the effects of ride-pooling, the granularity of the reward signals, and computational complexity. In addition, different applications may have specific considerations, such as safety, insurance.

11.8 Business Strategies

The research problems in the ridesharing domain are closely associated with how the ridesharing platforms run the operations. Innovation in product and business operations will continue to raise new challenging research problems. There can be multiple alternative product forms to achieve the same goals or address the same challenges, and they inherently define different optimization problems that RL can help tackle.

Surge pricing, for example, is a pricing strategy during the peak hours to address the severe shortage of supply with respect to the surging demand. We have explained its motivation in Chap. 4. While surge pricing is a common practice nowadays, it is not the only strategy that the ridesharing platforms adopt. Passenger requests can be queued if there are no vacant vehicles around to immediately serve the requests (Zhong et al. 2020). The queuing mechanism is perceived in some markets as a more socially acceptable mechanism during the peak hours than surge pricing. Several operational decision questions immediately come up, e.g., how large an area each queue should cover, if the coverage should be dynamically updated, and when the incoming requests should start queuing. These potentially time-varying decisions in a highly stochastic environment are good candidates to be solved for by RL.

Ridesharing platforms often use incentives to stimulate growth on both sides of the marketplace. The forms of incentives are diverse and ever evolving: rider coupons, discounts, target-based challenges, driver bonuses with spatial and temporal constraints, etc. Each incentive strategy changes the behaviors of a certain segment of the marketplace participants in a certain way, and they inevitably interact with the other marketplace levers, e.g., dynamic pricing (Yang et al. 2020b). The collective effects of the evolving incentives convolute the environment and dynamics of the marketplace, posing significant challenges to RL and other optimization methods. How to represent and capture these factors or explicitly model them in joint optimization is key to tackle these challenges.

Third-party service integrator allows passengers to simultaneously request orders from multiple ride-hailing platforms (Zhou et al. 2022). Service integrators offer the platforms more access to the demand but also bring competition more explicit by displaying the matching information (e.g., trip fare, pick-up distance) side by side. Optimizing pricing and matching policies in a competitive environment with feedback from the service integrator on the competition landscape will be interestingly different from those without a service integrator or in a non-competitive environment. With the added environment complexity, these problems are challenging to solve by traditional methods and could be better tackled by RL.

11.9   General RL

RL provides the necessary tools for the methods reviewed in this book. Hence, the problems of RL for ridesharing tie closely to the development in RL in general. In the context of ridesharing, we have seen from the literature review above that it is difficult for RL to learn combinatorial actions, e.g., the system matching actions. In the era of deep RL, model interpretability is a long-standing challenge, which hampers investigation of customer experience corner cases. For experience-critical service like ridesharing, policy exploration adds further complication, especially for real-world deployment. In view of these challenges, the future is probably that RL-based and traditional optimization approaches will be complementing each other for a long time. We have seen such combinations in the current literature as Xu et al. (2018), Qin et al. (2021b) for matching, Chaudhari et al. (2020a), Jiao et al. (2021) for repositioning, and Delarue et al. (2020) for VRP, that combine RL with combinatorial optimization, mixed-integer programming, and tree search. The breakthroughs of RL that we are seeing in other domains and the continued development of RL methodology for ridesharing certainly make it exciting to anticipate the future landscape.

Closing Remarks 12

The ridesharing system is a complex multi-agent system with multiple decision levers. RL offers a powerful modeling vehicle for optimizing this system, but as we have seen from the current literature, challenges remain in tackling complexity in the learning algorithms, the coordination among the agents, and the joint optimization of multiple levers. Along tackling these challenges, we expect that domain knowledge in ridesharing as well as transportation in general will be increasingly instrumental to the successful adoption of RL.

This book aims to provide a big picture of what underlies a ridesharing platform, how RL can be used to solve the critical decision problems there, and the open resources available to support research in this field. As one may have noticed, most of the literature (as well as our own works) have just appeared in the last few years at the time of writing, and we expect it to continue growing and updating rapidly. If you have any comments/suggestions, have found any errors in the book, or simply want to discuss some ideas, please feel free to contact zq2107@caa.columbia.edu.

© The Author(s), under exclusive license to Springer Nature Switzerland AG 2025   97
Z. (Tony). Qin et al., *Reinforcement Learning in the Ridesharing Marketplace*,
Synthesis Lectures on Learning, Networks, and Algorithms,
https://doi.org/10.1007/978-3-031-59640-7_12

Algorithms A

A.1 Pricing & Incentives

Algorithm A.1 InBEDE

1: **Inputs:**
    order list OL_0 at $t = 0$, idle driver list DL_0 at $t = 0$;
2: **Initialize:**
    θ, ϕ
3: **for** Iteration $1, \ldots,$ **do**
4:   **for** $t = 0, \ldots, T$ **do**
5:     Update OL_t and DL_t;
6:     **for** Request i that arrives at t **do**
7:       Observe feature x_i;
8:       Select an arm a_i according to the bandit algorithm θ;
9:       **if** Request i converts into order **then**
10:         Append request i into order list OL_t
11:       **end if**
12:       Perform dispatch among OL_t and DL_t through linear assignment, using the driver value network ϕ;
13:       **for** Request i that arrives within t **do**
14:         Get estimated reward $\hat{u}_\pi(x_i, a_i)$ using Eq. (4.1)
15:         Update parameter θ with $(x_i, a_i, \hat{u}_\pi(x_i, a_i))$;
16:       **end for**
17:     **end for**
18:   **end for**
19:   Collect driver trajectories (s, a', s', r);
20:   Update parameter ϕ with the collected driver trajectories;
21: **end for**
22: **return** θ, ϕ;

© The Editor(s) (if applicable) and The Author(s), under exclusive license to Springer Nature Switzerland AG 2025
Z. (Tony). Qin et al., *Reinforcement Learning in the Ridesharing Marketplace*,
Synthesis Lectures on Learning, Networks, and Algorithms,
https://doi.org/10.1007/978-3-031-59640-7

Algorithm A.2 POMEE

Input: $D_{real} = \{\tau_1, \tau_2, \ldots, \tau_n\}$: The observed real trajectories over T steps;
      N: Number of trajectories generated in each iteration;
      K: Steps of generator per discriminator step;
      I: Number of iterations;

1: Initialize parameters θ^{he} and θ^a of policy π_{he} and π_a, parameters σ of discriminator D;
2: **for** $i = 1, 2, \ldots, I$ **do**
3:    **for** $k = 1, 2, \ldots, K$ **do**
4:       $\tau_{sim} = \varnothing$;
5:       **for** $j = 1, 2, \ldots, N$ **do**
6:          $\tau_j = \varnothing$;
7:          Randomly sample a trajectory τ_r from D_{real} and set its first state as the initial state o_0^A ;
8:          **for** $t = 0, 1, 2, \ldots, T - 1$ **do**
9:             Simulate the action $a_t^A = \pi_a(o_t^A)$ and then the action $a_t^E = \pi_{he}(o_t^A, a_t^A)$;
10:            Get reward r_t^a and r_t^{he}, respectively, according to Eqs. (4.10) and (4.9);
11:            Get next observation o_{t+1}^A given o_t^A, a_t^E by predefined transition;
12:            Add $\{o_t^A, a_t^A, a_t^E, r_t^a, r_t^{he}\}$ to τ_j;
13:          **end for**
14:          Add τ_j to τ_{sim};
15:       **end for**
16:       TRPO update θ^a and θ^{he}, respectively, with τ_{sim} according to Eqs. (4.5) and (4.6);
17:    **end for**
18:    Update the parameters σ of discriminator D by minimizing the loss in Eqs. (4.7) and (4.8) in turn;
19: **end for**
20: **return** the trained environment policies π_e, π_h .

A.2　　Online Matching (Dispatching)

A.3　　Vehicle Repositioning

A.4　　Ride-Pooling (Carpool)

Algorithm A.3 Unified Value Learning Framework for Dynamic Order Dispatching and Driver Repositioning (**V1D3**)

1: Given: the ensemble weight $1 > \omega > 0$, the reposition threshold $C > 0$ (usually chosen between 150 and 300).
2: Given: the offline evaluated value function V_{ope}.
3: Compute the set $\mathcal{E}$ containing the changing time points to re-ensemble.
4: Initialize the state value network V with random weights θ.
5: **for** the dispatch round $t = 1, 2, \ldots, N$ **do**
6:　　**if** $t \in \mathcal{E}$ **then**
7:　　　　$\forall s, \ V_\theta(s) \leftarrow \omega V_\theta(s) + (1 - \omega)V_{ope}^t(s)$.
8:　　**end if**
9:　　Solve the dispatch (assignment) problem (5.2) given the current value V_θ.
10:　　**if** $t \bmod C = 0$ **then**
11:　　　　Collect all drivers with idle time exceeding C time steps.
12:　　　　Compute the destination distribution (5.16) for each driver given the current value V_θ.
13:　　　　Reposition each driver stochastically according to the distribution.
14:　　**end if**
15:　　Obtain the system state $\mathcal{D}_D$, $\mathcal{D}_I$ and $\mathcal{D} = \mathcal{D}_D \cup \mathcal{D}_I$.
16:　　Construct the gradient of the learning objective (5.15), i.e., $\nabla L(\mathcal{D}; \theta)$ based on the current system state $\mathcal{D}$.
17:　　Update the state value network by performing a gradient descent step on θ, e.g., $\theta \leftarrow \theta - \alpha \nabla L(\mathcal{D}; \theta)$
18: **end for**
19: **return** V

Algorithm A.4 Reinforcement Learning in the Wild (RLW) Procedure

---

1: **Initialize:**
   $V, S, LR, R^*, dV^*, P^*, feedback, th, Cancel\_Model, T_{up}$
2: **Inputs:**
   W_{rew}, W_p, γ, C
3: $Batch \leftarrow \{\}$
4: **for** $t = 0, 1, 2, .., T$ **do** $\triangleright$ t denotes rounds of dispatch
5:    $O_t, D_t, G_t \leftarrow$ observation(t)
6:    // Order-Driver Assignment
7:    $w_{rew} =$ Interpolate(W_{rew}, t)
8:    $w_p =$ Interpolate(W_p, t)
9:    $w_{res} = 1 - w_{rew}$
10:   $E, P_C = \{\}, \{\}$
11:   **for** od in G_t **do** $\triangleright$ od is an order-driver pair
12:      $o, d =$ order and driver for od
13:      $r =$ Standardize(R^*, S_o) $\triangleright$ S_o is smoothed reward
14:      $dv =$ Standardize$(dV^*, \gamma V_o - V_d)$
15:      $p =$ Standardize(P^*, o, d)
16:      $P_C[od] = Cancel\_Model(od)$
17:      **if** $P_C[od] > th$ **then**
18:         $E_{od} = P_C[od] \cdot (w_{rew} \cdot r + w_{res} \cdot dv - w_p \cdot p)$
19:      **end if**
20:   **end for**
21:   $A \leftarrow$ Hungarian Algorithm(E) Kuhn (1955)
22:   $Batch.append(A)$ $\triangleright$ A: list of batch assignments
23:   // State Value Updating
24:   **if** t mod $T_{up} == 0$ **then**
25:      **for** $type, item$ in $Batch$ **do**:
26:         **if** $type$ is dispatch **then**
27:            $o, d =$ order and driver for $item$
28:            $S \leftarrow$ Reward Smoothing(S, r_o) $\triangleright$ Equation 5.21
29:            $p_c = P_C[od]$
30:            $\delta = p_c \cdot (S_o + \gamma V_o - V_d) + (1 - p_c) \cdot (\gamma V_d - V_d)$
31:            $R^* \leftarrow$ Stdizer_update(R^*, S_o) $\triangleright$ Equation 5.23
32:            $dV^* \leftarrow$ Stdizer_update$(dV^*, \gamma V_o - V_d)$
33:            $P^* \leftarrow$ Stdizer_update(P^*, o, d)
34:         **else if** $type$ is idle **then**
35:            $\delta = \gamma V_d - V_d$
36:         **end if**
37:         $V_d, LR_d \leftarrow$ Adam Update(LR_d, δ)
38:      **end for**
39:   **end if**
40:   Return driver assignments based on A
41:   Update $feedback$ based on settled rider requests
42:   $CR_t, AR_t \leftarrow feedback$
43:   **if** t mod C $== 0$ **then**
44:      $th \leftarrow$ LM_UCB(CR_t, AR_t, th) $\triangleright$ Algorithm A.5
45:   **end if**
46: **end for**

Algorithm A.5 Limited-Memory UCB (LM-UCB)

Require: $c, \alpha, \gamma, Arms$

$\quad Q_i \leftarrow 0$ ▷ for $i \in$ Arms

$\quad N_i \leftarrow 0$ ▷ for $i \in$ Arms

$\quad th \leftarrow RandomChoice\{Arms\}$

$\quad n \leftarrow 0$

$\quad$ **for** $t = 0 : T$ **do**   ▷ t in seconds

$\quad\quad$ Maximum Bipartite Matching(th)

$\quad\quad$ Value Updating

$\quad\quad$ **if** $mod(t, 60) = 0$ **then**

$\quad\quad\quad n \leftarrow \gamma n + 1$

$\quad\quad\quad N_a \leftarrow \gamma N_a : \forall a \in Arms$

$\quad\quad\quad q \leftarrow CR + 0.1 \times AR$ ▷ tentative function

$\quad\quad\quad Q_{th} \leftarrow \alpha Q_{th} + (1 - \alpha)q$

$\quad\quad\quad N_{th} \leftarrow N_{th} + 1$

$\quad\quad\quad th \leftarrow argmax\{Q_a + c\sqrt{\frac{\log n}{N_a}} : \forall a \in Arms\}$

$\quad\quad$ **end if**

$\quad$ **end for**

Summary of References

B

In this appendix, we summarize the key aspects of each paper referenced in the main text of this book. They are grouped by the topics in the order that they have appeared in the previous sections.

B.1   Pricing & Incentives

(See Table B.1).

B.2   Online Matching

(See Tables B.2 and B.3).

B.3   Vehicle Repositioning

(See Tables B.4 and B.5).

B.4   Dynamic Routing

(See Tables B.6 and B.7).

B.5   VRP

(See Tables B.8 and B.9).

© The Editor(s) (if applicable) and The Author(s), under exclusive license to Springer Nature Switzerland AG 2025
Z. (Tony). Qin et al., *Reinforcement Learning in the Ridesharing Marketplace*,
Synthesis Lectures on Learning, Networks, and Algorithms,
https://doi.org/10.1007/978-3-031-59640-7

Table B.1 Summary of literature for pricing

Paper	Agent	State	Action	Reward	Algorithm	Environment
Wu et al. (2016)	Global decision-maker	Current trip price (same for all trips), SD info	Price	Profit	Q-learning	No spatiotemporal dimensions
Chen et al. (2019a)	Global decision-maker	Features of the trip request	Discretized price change percentage	Profit	Contextual bandits with action values partly computed by CVNet	Ride-hailing simulator with pricing module and passenger elasticity model
Turan et al. (2020)	Global decision-maker for pricing and EV charging	Electricity price in each zone, passenger queue length for each OD pair, number of vehicles in each zone and their energy levels	Price for each OD pair, reposition/charging for each vehicle	Trip revenue—penalty for queues—operational cost for charging and reposition	PPO	Simulator
Song et al. (2020)	Global decision-maker	Location, time	Price for spatialtemporal grid cells	Trip price minus penalty for driver waiting	Q-learning	Case study: ride-hailing simulation of Seoul
Mazumdar et al. (2017)	Passenger	Price multiplier, time, if a ride has completed	Wait, take current ride	Trip price to pay	Risk-sensitive inverse RL	Historical data
Chen et al. (2021a)	Global decision-maker	Number of open requests, vacant vehicles, and occupied vehicles in each grid cell at time t, and demand in time $t-1$	Joint actions of price (per-km for excess mileage) and wage (per-km rate) for each grid cell	Profit: revenue minus wage	PPO	Simulation based on Hangzhou data from DiDi; modeling on both supply and demand elasticity

Table B.2 Summary of literature for online matching

Paper	Agent	State	Action	Reward	Algorithm	Environment	Notes
Xu et al. (2018)	Driver	Assignment to a specific order or idle	Location, time	Trip price	Tabular TD(0) for learning state values offline + Hungarian method for generating the assignment online	Deployed in production; multi-agent, agent-level simulation	
Wang et al. (2018)	Driver	Assignment to a specific order or idle	Location, time, SD features	Trip price	Offline DQN for matching, CFPT network for transfer learning	Single-agent simulation	Single-vehicle problem
Tang et al. (2019), Qin et al. (2020a)	Driver	Assignment to a specific order or idle	Location, time, SD features	Trip price	CVNet (deep TD-like algorithm) for learning state values offline + Hungarian method for generating the assignment online	Deployed in production; multi-agent, agent-level simulation	Hierarchical sparse coarse coding, cerebellar embedding of spatial info, Lipschitz regularization on network
Holler et al. (2019)	Driver, system	Matching a driver to an order, reposition a driver; matching and repositioning done sequentially	Global info of all drivers and open orders	Trip price, reposition cost	DQN, PPO	Multi-agent, agent-level simulation	Attention mechanism to extract global state info into a context vector
Li et al. (2019)	Driver	Assignment to a specific order or idle	Location, time, is_available	Trip price	Mean-field MARL, AC method	Homogeneous vehicles within the same grid cell	The mean action is represented by the peers' destination distribution

(continued)

Table B.2 (continued)

Paper	Agent	State	Action	Reward	Algorithm	Environment	Notes
Jin et al. (2019)	Worker: hex cell manager: group of hex cells (one layer)	Worker: ranking for match and reposition manager: abstract goal for workers	Number of vehicles, orders, entropy, reposition-guided vehicles, distributions of trip prices and durations in the given hex cell	Manager: total driver income + specifically designed quantity to promote high order response rate worker: intrinsic reward for following the goal generated by manager	Hierarchical MARL	Homogeneous vehicles within the same grid cell	Multiple managers, each manager communicates with multiple workers multi-head attention mechanism for coordination
Zhou et al. (2019)	Driver	A trip tuple: (origin cell, destination cell, trip duration, price)	Cell index, number of idle vehicles, orders, distribution of trip destinations in the given hex cell	Trip price	Independent learning with KL divergence regularization	Homogeneous vehicles within the same grid cell	
Ke et al. (2020b)	Trip request	Match or delay	Global: number of idle vehicles, open requests, expected arrival rates of requests and drivers in each cell local: location, cumulative waiting time, expected pick-up distance	Local reward based on the ultimate outcome of matching (whether or not matched or cancelled): trip value, pick-up distance, and match window time Global reward is based on average local reward. Final reward is convex combination of local and global rewards. The rewards have to be updated at the end of the epoch	DQN, PPO, A2C, ACER with delayed reward. Whole episode trajectories are sampled from replay buffer	Agent-based simulation with delayed matching feature	

Table B.3 Summary of literature for online matching

Paper	Agent	State	Action	Reward	Algorithm	Environment	Notes
Wang et al. (2019)	System	Expected length of current batch	Current nodes in the bipartite graph, current batch size	The sum of the edge weights in the batch	Restricted Q-learning	DiDi GAIA data set	The adaptive batch-based matching has a guarantee on competitive ratio
Qin et al. (2021a)	System	Match the current batch or continue to batch (decision made at every time interval)	Number of batched requests and vehicles and estimated arrival rates of demand and supply in each cell	For each time interval, the negative of total matching wait time for all batched requests and the total pick-up wait time saved (by delaying the current batch)	ACER that combines on-policy updates (through a queuing-based simulator) with off-policy updates	Shanghai taxi data	
Shi et al. (2019a)	Vehicle	Remaining battery level when available, next available time and location, global time	Matching, EV (charging)	Trip price—pick-up cost—charging cost	Similar to CVNet Tang et al. (2019)	Synthetic data	Assumes decomposibility of system value into vehicle values
Kullman et al. (2022)	System	Global time, new request info, state of each vehicle	Joint matching, reposition, and charging (EV) for each vehicle	Revenue—travel cost	DQN + attention mechanism over vehicles embeddings (similar to Holler et al. (2019))	NYC taxi data with taxi zones	Decision epoch is either a new request arrives or a vehicle becomes idle
Al-Kanj et al. (2020)	System	Supply-demand counts in spatiotemporal discretized space	Joint matching and reposition, EV (charging)	Revenue—charging expense	ADP with value function approx. on post-decision states. Value function approx. by hierarchical aggregation	Simulation with New Jersey ride-hailing trip data	

Table B.4 Summary of literature for vehicle repositioning (taxi routing)

Paper	Type	Agent	State (in addition to ST info)	Action	Reward	Episode	Algorithm	Coordination	Data
Rong et al. (2016)	Taxi	Driver	Direction from which driver arrives at the current location	Neighboring cells in a grid system	Trip fare	Long-term	Model-based, VI to solve MDP	–	
Han et al. (2016)	Taxi	Driver	–	Neighboring cells in a grid system	Trip fare—reposition cost	From idle to completion of next trip(s) and being idle again	Q-learning	–	
Verma et al. (2017)	Taxi	Driver	–	Neighboring cells in a grid system	Trip fare—reposition cost	Up to the next match	MC learning	–	Taxi log data from Singapore
Wen et al. (2017)	Taxi	Driver	SD contextual info	Neighboring cells in a grid system	Saved idle time compared to a counterfactual simulation without reposition. Penalized if no match after reposition	Up to the next match	DQN with greedy action Compared with MINLP and SAR (simple anticipatory rebalancing) on avg passenger wait time	–	
Garg and Ranu (2018)	Taxi	Driver	Current node on road network	Adjacent node or edge	Idle cruising distance till the next passenger	Up to the next match	MAB + MCTS	–	
Gao et al. (2018)	Taxi	Driver	Location, occupied, parking, vacant	Idle cruise to a neighboring cell in the grid system, carrying passenger to destination, waiting	Trip mileage/idle cruising mileage Problem objective: effective driving ratio = total trip mileage/total idle cruising mileage	Long-term (day)	Q-learning	–	Beijing taxi data 2013

(continued)

Table B.4 (continued)

Paper	Type	Agent	State (in addition to ST info)	Action	Reward	Episode	Algorithm	Coordination	Data
Zhou et al. (2018)	Taxi	Driver	Current road segment, time, previous road segment	Adjacent node or edge	Trip fare	Long-term	Model-based, VI to solve MDP	–	A year of taxi data from a major city in China
Yu and Shen (2019)	Taxi	Driver	Current node on road network	Adjacent node or edge	Trip fare—operational cost	Long-term (day)	Model-based, VI to solve MDP; use parallel matrix operations to accelerate computation	–	Shanghai taxi trajectory data in the morning of a weekday
Shou et al. (2020)	Taxi, system	Driver	Boolean: whether or not assigned to consecutive requests	Neighboring cells in a grid system	Trip fare—operational cost operational cost per unit distance learned through IRL	Long-term (day)	Model-based, VI to solve MDP; inverse RL to learn unit distance op cost	Sequentially make decision for each driver. Adjust the matching prob of each driver's MDP, and solve again	Beijing ride-hailing trajectory data from DiDi over three weekdays
Jiao et al. (2021)	Taxi	Vehicle	SD contextual info in the current cell	Neighboring cells in a grid system	Trip fare—reposition cost	Long-term	Offline CVNet + decision-time action search	–	DiDi ride-hailing data

Table B.5 Summary of literature for vehicle repositioning (system reposition)

Paper	Type	Agent	State (in addition to ST info)	Action	Reward	Episode	Algorithm	Coordination	Data
Lin et al. (2018)	System	Driver	Global SD contextual features in all cells	Neighboring cells in a grid system	Trip fare, shared when multiple agents are in the same grid cell	Long-term	Contextual DQN, AC	Contextual state features, action space pruned by context	4 weeks of DiDi data in Chengdu, China
Oda and Joe-Wong (2018)	System	Driver	Global SD state discretized into cells, treated as an image	Reachable cells in a grid system within the reposition cycle	Weighted number of pick-ups—reposition time	Long-term	Independent DQN	–	NYC taxi data
Shou and Di (2020b)	System	Vehicle	–	Neighboring cells in a grid system	Trip fare	Long-term	Bilevel optimization: top Bayesian optimization to update reward param, bottom mean-field MARL (AC)	Mean-field MARL	NYC taxi data
Jiao et al. (2021)	System	Vehicle	SD contextual info in the current and neighboring cells	Neighboring cells in a grid system	Trip fare—reposition cost	Long-term	Deep SARSA	Stochastic policy through softmax of action values, SD regularization to action values	DiDi ride-hailing data
Mao et al. (2020)	System	System	SD info for each zone	Reposition plan: number of repositioned vehicles for each OD pair in a zone map	Monetized passenger waiting time	Long-term	Batch AC	Central decision-making	NYC taxi data

(continued)

Table B.5 (continued)

Paper	Type	Agent	State (in addition to ST info)	Action	Reward	Episode	Algorithm	Coordination	Data
Feng et al. (2020)	System	System	Status of every vehicle and request	Atomic action: driver-passenger match or driver-destination match (reposition); system action: sequence of atomic actions	Trip fare—operational cost	Long-term (day)	PPO applied to MDP with sequential decision process embedded: global actions decomposed into sequential atomic ones	Central decision-making	DiDi data: 5 regions, 1000 cars, 360 min
Liu et al. (2020)	System	Vehicle	Discrete zone structure constructed by clustering a road connectivity graph	Neighboring zones	Trip fare	Long-term	Contextual DQN with shared value function	Vehicle actions generated sequentially	Real-world taxi data
Zhang et al. (2020b)	System	Vehicle	Global SD distributions	Neighboring cells in a grid system	Global KL distance between SD distributions	Long-term	DQN + Q-table in tandem with shared value function	Global state features	DiDi data
Chaudhari et al. (2020a)	System	Vehicle	Cell in an ST table	Neighboring cells in a grid system, wait	Trip fare—operational cost	Long-term	SARSA-like policy evaluation	Solve an assignment problem of surplus and deficits in terms of SD gap	NYC taxi data

Table B.6 Summary of literature for route guidance

Paper	Type	Equilibrium	Agent	State	Action	Reward	Algorithm	Road network
Mainali et al. (2008)	TAP	UE	Vehicle	Current node (intersection)	An adjacent link	Travel time on the link	Q-iteration. The route is constructed by following the decision at each intersection	Grid network
Bazzan and Chira (2015)	TAP	SE	Vehicle	–	Route from feasible routes for the OD pair	Total travel time on the chosen route	Hybrid method: Q learning for individual agent + Genetic algorithm for system equilibrium, minimizing avg travel time over different trips	–
Ramos et al. (2018)	TAP	UE (empirical convergence)	Vehicle	–	Route from feasible routes for the OD pair	Total travel time on the chosen route	Q-learning with action-regret updates to minimize driver's total regret	Braess graphs, OW network
Zhou et al. (2020a)	TAP	UE (theoretical convergence established)	Vehicle	–	Route from feasible routes for the OD pair	Total travel time on the chosen route	B-M RL scheme, similar to a MARL algo with individual reward	Nguyen-Dupuis network
Kim et al. (2005)	DR	UE	Vehicle	Node, time, binary congestion status vector for all links	An adjacent node	Cost accrued by traversing the link	Parameters of MDP estimated from data. MDP solved by value iterations	Southeast Michigan network and traffic data

(continued)

Table B.6 (continued)

Paper	Type	Equilibrium	Agent	State	Action	Reward	Algorithm	Road network
Turner et al. (2008)	DR	SE	Vehicle	Just a single link	Start time on the link	Difference reward (uplift): with and without the agent in the system	Q-learning	–
Yu et al. (2012)	DR	UE	Vehicle	Current node	An adjacent link; action instruction received from the intersection in real time	Travel time on the link	Similar approach to Mainali et al. (2008), but incremental update is done by a step of SARSA. The updates are done in real time and value functions are sync-ed at each intersection, which is an independent traffic management module	SOUND/4U simulator based on the road network of Kurosaki, Kitakyushu in Japan
Grunitzki et al. (2014)	DR	SE	Vehicle	Current node	An adjacent link	Individual reward: negative travel time experienced by the agent on a link difference reward: as in Turner et al. (2008)	Applied to a more sophisticated network than aamas08 paper. Results show that DQ-learning outperforms IQ-learning	Abstract network topology

Table B.7 Summary of literature for route guidance

Paper	Type	Equilibrium	Agent	State	Action	Reward	Algorithm	Road network
Mao and Shen (2018)	DR	UE	Vehicle	Current node, time, discrete congestion states vector of the arcs at most 2 steps away	An adjacent node	Negative travel time experienced by the agent on a link	Offline batch RL: fitted Q-iterations with tree-based function approximator (Extreme Randomized Trees). Compared with model-based Q-iterations	Sioux falls network
Bazzan and Grunitzki (2016)	DR	UE	Trip (OD pair)	Current node	An adjacent link	Negative travel time experienced by the agent on a link	Independent Q learning compared with successive average method	OW network, Sioux Falls network
Wen et al. (2019)	DR	UE	Vehicle	Next approaching node, destination node	An adjacent link	Negative travel time experienced by the agent on a link	Tabular Q-learning, global road network clustered into subnetworks by differential evolution-based clustering Top-level network contains only boundary nodes of the original network. Top network policy produces the 'destination' node for a subnetwork. Subnetwork policy provides link-level guidance to reach its sub-destination	SUMO simulator with various networks in Japan and US
Shou and Di (2020b)	DR	SE (avg travel time), UE (travel distance)	Vehicle	Current node, time	An adjacent link	Negative average travel time of all agents or travel distance on a link	Bilevel optimization: Lower level—mean field MARL to solve for dynamic routing for travelers Upper level—Bayesian optimization to optimize controls by city planners	SUMO with Manhattan network

Table B.8 Summary of literature for VRP

Paper	Type	State	Action	Reward	Network	Algorithm	Problem size
Nazari et al. (2018)	Single-vehicle	Location and demand of each request	The next request to visit	Negative travel distance	Non-sequential encoder for the input with RNN decoder that attends over the input space (Pointer network without an RNN encoder)	AC	Single-vehicle capacitated VRP with split delivery: one active vehicle at a time
Kool et al. (2018)	Single-vehicle	Coordinates and original demand of each node, remaining demand of each node, remaining capacity of the vehicle	Next stop for a given vehicle	Negative travel distance	Transformer encoder with input of coordinates and demand of each node + self-attention-based decoder with additional input of remaining demands and capacity at time t	REINFORCE with greedy rollout baseline	TSP and Capacitated VRP with split delivery, 100 nodes
Balaji et al. (2019)	Single-vehicle	Current pickup location, vehicle's location and remaining capacity, orders' locations, statuses, waiting times, and values	Accept an order, pick up an accepted order, wait	Value of delivered order (accept, pickup, deliver)—cost (waiting, traveling, penalty)	Two dense layers NN	APE-X DQN Horgan et al. (2018), a variant of a DQN that utilizes distributed prioritized experience replay	Stochastic and dynamic CVRP with pick-up/delivery and time windows, 8 × 8 map, 5 orders 3 pick-up locations
James et al. (2019)	Multi-vehicle	System state (available requests, charging station output, vehicles' status (location, battery levels, next stops), vehicle tour graph	Next stop for a given vehicle; The vehicles generate actions sequentially at each time step	Expected objective value for a tour	Pointer network for actor, another critic network; Network architecture is similar to NeurIPS-18 paper, but with structural graph embedding (Struct2Vec) for the encoder	A3C	Multi-vehicle dynamic VRP with pick-up and delivery for EVs: 200 random requests, 100 vehicles

(continued)

Table B.8 (continued)

Paper	Type	State	Action	Reward	Network	Algorithm	Problem size
Zhang et al. (2020a)	Multi-vehicle	Same as Kool et al. (2018)	Same as Kool et al. (2018); The vehicles generate actions sequentially at each time step	Negative travel distance + negative constraint violation penalty	Same as Kool et al. (2018)	Same as Kool et al. (2018)	Multi-vehicle VRP with soft time windows (no split delivery): 150 nodes, 5 vehicles
Ulmer et al. (2020)	Single-vehicle	Vehicle location, time, num of passengers onboard, info of in-process and outstanding requests, route plan from last epoch	The next stop to visit and the new route plan; The paper defines a new variant of MDP called route-based MDP	Difference in route value between the old route plan and the new one	N.A.	Insert new request s into the current route and use variable neighborhood search to improve the route	SDVRP with pick-up and delivery (DDARP)
Joe and Lau (2020)	Multi-vehicle	Includes the cost for the remaining route for each vehicle	Matching a new order to a vehicle. Rerouting after matching is done by simulated annealing for VRP	Cost diff between two consecutive decisions	Not specified	NN-based TD learning with experience replay (like in Tang et al. (2019)) to learn the action-value function	Multi-vehicle dynamic VRP with pick-up/delivery and delivery windows: 48 nodes, 2 vehicles, avg 22 orders/day

Table B.9 Summary of literature for VRP

Paper	Type	State	Action	Reward	Network	Algorithm	Problem size
Delarue et al. (2020)	Single-vehicle	The remaining unvisited nodes	To generate one route (tour) through solving a prize collecting TSP by MIP	Negative tour distance	Value network consists of dense layers + ReLU activation (representable by mixed-integer linear constraints)	MC policy iteration: rollout N trajectories, fit a new NN	CVRP: 51 nodes
Duan et al. (2020)	Single-vehicle	Nodes (location, demand), edges (distance, adjacency)	Generate one node at a time sequentially; The resulting sequence may have multiple depot occurrences for different tours	Negative travel distance	GCN-based encoder with both node and edge features; GRU-based decoder similar to the pointer network as policy network and MLP-based decoder on the edge encoding as classifier	REINFORCE with greedy rollout baseline Kool et al. (2018) to train the policy network; Cross-entropy loss to train the binary classifier of route edges with policy output as labels	CVRP: 400 nodes
Lin et al. (2021)	Multi-vehicle	For time t, the state of each vertex (location, time window, remaining demand), and global variables (time, battery level of the active vehicle, number of EVs not in charging)	Next stop for the current route; Unlike James et al. (2019) the routes of the vehicles are generated sequentially. Every time the depot appears in the sequence, the system time is reset to 0	Negative travel distance + negative penalties for constraint violations	Encoder with 1D conv layer and graph embedding for the nodes and attention layer; LSTM-based decoder	REINFORCE with greedy rollout baseline Kool et al. (2018)	EV with time window and charging. Within the planning horizon, a vehicle can visit the depot only once: C100, S12, EV12

References

Abe, N., Melville, P., Pendus, C., Reddy, C. K., Jensen, D. L., Thomas, V. P., Bennett, J. J., Anderson, G. F., Cooley, B. R., Kowalczyk, M., et al. (2010). Optimizing debt collections using constrained reinforcement learning. In *Proceedings of the 16th ACM SIGKDD international conference on Knowledge discovery and data mining*, pages 75–84.

Abe, N., Verma, N., Apte, C., and Schroko, R. (2004). Cross channel optimized marketing by reinforcement learning. In *Proceedings of the tenth ACM SIGKDD international conference on Knowledge discovery and data mining*, pages 767–772.

AboElHamd, E., Shamma, H. M., and Saleh, M. (2020). Dynamic programming models for maximizing customer lifetime value: an overview. In *Intelligent Systems and Applications: Proceedings of the 2019 Intelligent Systems Conference (IntelliSys) Volume 1*, pages 419–445. Springer.

Al-Kanj, L., Nascimento, J., and Powell, W. B. (2020). Approximate dynamic programming for planning a ride-hailing system using autonomous fleets of electric vehicles. *European Journal of Operational Research*, 284(3):1088–1106.

Alabbasi, A., Ghosh, A., and Aggarwal, V. (2019). Deeppool: Distributed model-free algorithm for ride-sharing using deep reinforcement learning. *IEEE Transactions on Intelligent Transportation Systems*, 20(12):4714–4727.

Albus, J. S. (1971). A theory of cerebellar function. *Mathematical Biosciences*, 10(1-2):25–61.

Alonso-Mora, J., Samaranayake, S., Wallar, A., Frazzoli, E., and Rus, D. (2017a). On-demand high-capacity ride-sharing via dynamic trip-vehicle assignment. *Proceedings of the National Academy of Sciences*, 114(3):462–467.

Alonso-Mora, J., Wallar, A., and Rus, D. (2017b). Predictive routing for autonomous mobility-on-demand systems with ride-sharing. In *2017 IEEE/RSJ International Conference on Intelligent Robots and Systems (IROS)*, pages 3583–3590. IEEE.

Angrist, J. D., Caldwell, S., and Hall, J. V. (2021). Uber versus taxi: A driver's eye view. *American Economic Journal: Applied Economics*, 13(3):272–308.

Arora, S. and Doshi, P. (2021). A survey of inverse reinforcement learning: Challenges, methods and progress. *Artificial Intelligence*, page 103500.

Z. (Tony). Qin et al., *Reinforcement Learning in the Ridesharing Marketplace*, Synthesis Lectures on Learning, Networks, and Algorithms, https://doi.org/10.1007/978-3-031-59640-7

Azagirre, X., Balwally, A., Candeli, G., Chamandy, N., Han, B., King, A., Lee, H., Loncaric, M., Martin, S., Narasiman, V., et al. (2023). A better match for drivers and riders: Reinforcement learning at lyft. *arXiv preprint* arXiv:2310.13810.

Baird III, L. C. (1993). Advantage updating. Technical report, WRIGHT LAB WRIGHT-PATTERSON AFB OH.

Balaji, B., Bell-Masterson, J., Bilgin, E., Damianou, A., Garcia, P. M., Jain, A., Luo, R., Maggiar, A., Narayanaswamy, B., and Ye, C. (2019). Orl: Reinforcement learning benchmarks for online stochastic optimization problems. *arXiv preprint* arXiv:1911.10641.

Bazzan, A. L. and Chira, C. (2015). A hybrid evolutionary and multiagent reinforcement learning approach to accelerate the computation of traffic assignment. In *AAMAS*, pages 1723–1724.

Bazzan, A. L. and Grunitzki, R. (2016). A multiagent reinforcement learning approach to en-route trip building. In *2016 International Joint Conference on Neural Networks (IJCNN)*, pages 5288–5295. IEEE.

Bei, X. and Zhang, S. (2018). Algorithms for trip-vehicle assignment in ride-sharing. In *Thirty-second AAAI conference on artificial intelligence*.

Beirigo, B. A., Negenborn, R. R., Alonso-Mora, J., and Schulte, F. (2022). A business class for autonomous mobility-on-demand: Modeling service quality contracts in dynamic ridesharing systems. *Transportation Research Part C: Emerging Technologies*, 136:103520.

Bello, I., Pham, H., Le, Q. V., Norouzi, M., and Bengio, S. (2016). Neural combinatorial optimization with reinforcement learning. *arXiv preprint* arXiv:1611.09940.

Berner, C., Brockman, G., Chan, B., Cheung, V., Dębiak, P., Dennison, C., Farhi, D., Fischer, Q., Hashme, S., Hesse, C., et al. (2019). Dota 2 with large scale deep reinforcement learning. *arXiv preprint* arXiv:1912.06680.

Bertsimas, D. and Perakis, G. (2006). Dynamic pricing: A learning approach. In *Mathematical and computational models for congestion charging*, pages 45–79. Springer.

Bimpikis, K., Candogan, O., and Saban, D. (2019). Spatial pricing in ride-sharing networks. *Operations Research*, 67(3):744–769.

Bojinov, I., Simchi-Levi, D., and Zhao, J. (2023). Design and analysis of switchback experiments. *Management Science*, 69(7):3759–3777.

Boutilier, C., Cohen, A., Hassidim, A., Mansour, Y., Meshi, O., Mladenov, M., and Schuurmans, D. (2018). Planning and learning with stochastic action sets. In *Proceedings of the 27th International Joint Conference on Artificial Intelligence*, pages 4674–4682. Accessed April 21, 2020.

Brandfonbrener, D., Whitney, W., Ranganath, R., and Bruna, J. (2021). Offline rl without off-policy evaluation. *Advances in Neural Information Processing Systems*, 34:4933–4946.

Braverman, A., Dai, J. G., Liu, X., and Ying, L. (2019). Empty-car routing in ridesharing systems. *Operations Research*, 67(5):1437–1452.

Brodsky, I. (2018). H3: Uber?s hexagonal hierarchical spatial index. available online: https://eng.uber.com/ h3/ (accessed on june 26, 2019).

Brown, E. (2020). *The Ride-Hail Utopia That Got Stuck in Traffic - WSJ.* https://www.wsj.com/articles/the-ride-hail-utopia-that-got-stuck-in-traffic-11581742802.

Buşoniu, L., Babuška, R., and Schutter, B. D. (2010). Multi-agent reinforcement learning: An overview. *Innovations in multi-agent systems and applications-1*, pages 183–221.

Chaudhari, H. A., Byers, J. W., and Terzi, E. (2020a). Learn to earn: Enabling coordination within a ride hailing fleet. *Proceedings of IEEE International Conference on Big Data*.

Chaudhari, H. A., Byers, J. W., and Terzi, E. (2020b). Simulation code for "learn to earn: Enabling coordination within a ride-hailing fleet". "https://transparent-framework.github.io/optimize-ride-sharing-earnings/".

Chen, C., Yao, F., Mo, D., Zhu, J., and Chen, X. M. (2021a). Spatial-temporal pricing for ride-sourcing platform with reinforcement learning. *Transportation Research Part C: Emerging Technologies*, 130:103272.

Chen, H., Jiao, Y., Qin, Z., Tang, X., Li, H., An, B., Zhu, H., and Ye, J. (2019a). Inbede: Integrating contextual bandit with td learning for joint pricing and dispatch of ride-hailing platforms. In *2019 IEEE International Conference on Data Mining (ICDM)*, pages 61–70. IEEE.

Chen, M. K. and Sheldon, M. (2016). Dynamic pricing in a labor market: Surge pricing and flexible work on the uber platform. *Ec*, 16:455.

Chen, X.-H., Luo, F.-M., Yu, Y., Li, Q., Qin, Z., Shang, W., and Ye, J. (2023). Offline model-based adaptable policy learning for decision-making in out-of-support regions. *IEEE Transactions on Pattern Analysis and Machine Intelligence*.

Chen, X.-H., Yu, Y., Li, Q., Luo, F.-M., Qin, Z., Shang, W., and Ye, J. (2021b). Offline model-based adaptable policy learning. *Advances in Neural Information Processing Systems*, 34:8432–8443.

Chen, Y., Qian, Y., Yao, Y., Wu, Z., Li, R., Zhou, Y., Hu, H., and Xu, Y. (2019b). Can sophisticated dispatching strategy acquired by reinforcement learning?-a case study in dynamic courier dispatching system. *arXiv preprint* arXiv:1903.02716.

Cheng, S.-F., Jha, S. S., and Rajendram, R. (2018). Taxis strike back: A field trial of the driver guidance system. In *AAMAS'18: Proceedings of the 17th International Conference on Autonomous Agents and Multiagent Systems, Stockholm, July 10*, volume 15, pages 577–584.

Chin, A. and Qin, Z. T. (2023). A unified representation framework for rideshare marketplace equilibrium and efficiency. In *Proceedings of the 31st ACM International Conference on Advances in Geographic Information Systems*, SIGSPATIAL '23, New York, NY, USA. Association for Computing Machinery.

Dantzig, G. B. and Ramser, J. H. (1959). The truck dispatching problem. *Management science*, 6(1):80–91.

Delarue, A., Anderson, R., and Tjandraatmadja, C. (2020). Reinforcement learning with combinatorial actions: An application to vehicle routing. *Advances in Neural Information Processing Systems*, 33:609–620.

Deng, Z., Jiang, J., Long, G., and Zhang, C. (2023). Causal reinforcement learning: A survey. *arXiv preprint* arXiv:2307.01452.

DiDi (2021). Didi decision intelligence simulation platform. "https://outreach.didichuxing.com/Simulation".

Dong, Y., Zhang, S., Liu, X., Zhang, Y., and Shen, T. (2021). Variance aware reward smoothing for deep reinforcement learning. *Neurocomputing*, 458:327–335.

Donovan, B. and Work, D. (2016). New york city taxi trip data (2010–2013).

Dosovitskiy, A., Ros, G., Codevilla, F., Lopez, A., and Koltun, V. (2017). Carla: An open urban driving simulator. In *Conference on robot learning*, pages 1–16. PMLR.

Duan, L., Zhan, Y., Hu, H., Gong, Y., Wei, J., Zhang, X., and Xu, Y. (2020). Efficiently solving the practical vehicle routing problem: A novel joint learning approach. In *Proceedings of the 26th ACM SIGKDD International Conference on Knowledge Discovery & Data Mining*, pages 3054–3063.

Eshkevari, S. S., Tang, X., Qin, Z., Mei, J., Zhang, C., Meng, Q., and Xu, J. (2022). Reinforcement learning in the wild: Scalable rl dispatching algorithm deployed in ridehailing marketplace. In *Proceedings of the 26th ACM SIGKDD International Conference on Knowledge Discovery & Data Mining*.

Farias, V., Li, A., Peng, T., and Zheng, A. (2022). Markovian interference in experiments. *Advances in Neural Information Processing Systems*, 35:535–549.

Feng, J., Gluzman, M., and Dai, J. (2020). Scalable deep reinforcement learning for ride-hailing. *IEEE Control Systems Letters*.

Finn, C., Christiano, P., Abbeel, P., and Levine, S. (2016). A connection between generative adversarial networks, inverse reinforcement learning, and energy-based models. *arXiv preprint* arXiv:1611.03852.

FortuneBusinessInsights (2021). *Rideshare Market Size, 2021-2028*. https://www.fortunebusinessinsights.com/ride-sharing-market-103336.

Gao, Y., Jiang, D., and Xu, Y. (2018). Optimize taxi driving strategies based on reinforcement learning. *International Journal of Geographical Information Science*, 32(8):1677–1696.

Garg, N. and Ranu, S. (2018). Route recommendations for idle taxi drivers: Find me the shortest route to a customer! In *Proceedings of the 24th ACM SIGKDD International Conference on Knowledge Discovery & Data Mining*, pages 1425–1434.

Gasse, M., Grasset, D., Gaudron, G., and Oudeyer, P.-Y. (2021). Causal reinforcement learning using observational and interventional data. *arXiv preprint* arXiv:2106.14421.

Grunitzki, R., de Oliveira Ramos, G., and Bazzan, A. L. C. (2014). Individual versus difference rewards on reinforcement learning for route choice. In *2014 Brazilian Conference on Intelligent Systems*, pages 253–258. IEEE.

Guériau, M., Cugurullo, F., Acheampong, R. A., and Dusparic, I. (2020). Shared autonomous mobility on demand: A learning-based approach and its performance in the presence of traffic congestion. *IEEE Intelligent Transportation Systems Magazine*, 12(4):208–218.

Guériau, M. and Dusparic, I. (2018). Samod: Shared autonomous mobility-on-demand using decentralized reinforcement learning. In *2018 21st International Conference on Intelligent Transportation Systems (ITSC)*, pages 1558–1563. IEEE.

Hales, T. C. (2001). The honeycomb conjecture. *Discrete & Computational Geometry*, 25(1):1–22.

Haliem, M., Mani, G., Aggarwal, V., and Bhargava, B. (2020). A distributed model-free ride-sharing algorithm with pricing using deep reinforcement learning. In *Computer Science in Cars Symposium*, pages 1–10.

Haliem, M., Mani, G., Aggarwal, V., and Bhargava, B. (2021). A distributed model-free ride-sharing approach for joint matching, pricing, and dispatching using deep reinforcement learning. *IEEE Transactions on Intelligent Transportation Systems*, 22(12):7931–7942.

Hall, R. W. (1986). The fastest path through a network with random time-dependent travel times. *Transportation science*, 20(3):182–188.

Hildebrandt, F. D., Thomas, B. W., and Ulmer, M. W. (2023). Opportunities for reinforcement learning in stochastic dynamic vehicle routing. *Computers & operations research*, 150, 106071.

Han, B., Lee, H., and Martin, S. (2022). Real-time rideshare driver supply values using online reinforcement learning. In *Proceedings of the 28th ACM SIGKDD Conference on Knowledge Discovery and Data Mining*, pages 2968–2976.

Han, M., Senellart, P., Bressan, S., and Wu, H. (2016). Routing an autonomous taxi with reinforcement learning. In *Proceedings of the 25th ACM International on Conference on Information and Knowledge Management*, pages 2421–2424.

Hao, X., Peng, Z., Ma, Y., Wang, G., Jin, J., Hao, J., Chen, S., Bai, R., Xie, M., Xu, M., et al. (2020). Dynamic knapsack optimization towards efficient multi-channel sequential advertising. In *International Conference on Machine Learning*, pages 4060–4070. PMLR.

Haydari, A. and Yilmaz, Y. (2020). Deep reinforcement learning for intelligent transportation systems: A survey. *IEEE Transactions on Intelligent Transportation Systems*.

Hinton, G., Vinyals, O., and Dean, J. (2015). Distilling the knowledge in a neural network. In *NIPS Deep Learning and Representation Learning Workshop*.

Hinton, G. E. and Salakhutdinov, R. R. (2006). Reducing the dimensionality of data with neural networks. *science*, 313(5786):504–507.

Ho, J. and Ermon, S. (2016). Generative adversarial imitation learning. *Advances in neural information processing systems*, 29.

Holler, J., Vuorio, R., Qin, Z., Tang, X., Jiao, Y., Jin, T., Singh, S., Wang, C., and Ye, J. (2019). Deep reinforcement learning for multi-driver vehicle dispatching and repositioning problem. In Wang, J., Shim, K., and Wu, X., editors, *2019 IEEE International Conference on Data Mining (ICDM)*, pages 1090–1095. Institute of Electrical and Electronics Engineers, Washington, DC.

Horgan, D., Quan, J., Budden, D., Barth-Maron, G., Hessel, M., Van Hasselt, H., and Silver, D. (2018). Distributed prioritized experience replay. *arXiv preprint* arXiv:1803.00933.

Hu, B., Hu, M., and Zhu, H. (2022). Surge pricing and two-sided temporal responses in ride hailing. *Manufacturing & Service Operations Management*, 24(1):91–109.

Hu, M. and Zhou, Y. (2022). Dynamic type matching. *Manufacturing & Service Operations Management*, 24(1):125–142.

Huang, T., Li, Q., and Qin, Z. T. (2022). Multiple tiered treatments optimization with causal inference on response distribution. In *2022 IEEE International Conference on Big Data*. IEEE.

Iglesias, R., Rossi, F., Wang, K., Hallac, D., Leskovec, J., and Pavone, M. (2018). Data-driven model predictive control of autonomous mobility-on-demand systems. In *2018 IEEE International Conference on Robotics and Automation (ICRA)*, pages 1–7. IEEE.

James, J., Yu, W., and Gu, J. (2019). Online vehicle routing with neural combinatorial optimization and deep reinforcement learning. *IEEE Transactions on Intelligent Transportation Systems*, 20(10):3806–3817.

Jiao, Y., Tang, X., Qin, Z. T., Li, S., Zhang, F., Zhu, H., and Ye, J. (2021). Real-world ride-hailing vehicle repositioning using deep reinforcement learning. *Transportation Research Part C: Emerging Technologies*, 130:103289.

Jin, J., Zhou, M., Zhang, W., Li, M., Guo, Z., Qin, Z., Jiao, Y., Tang, X., Wang, C., Wang, J., et al. (2019). Coride: Joint order dispatching and fleet management for multi-scale ride-hailing platforms. In *Proceedings of the 28th ACM International Conference on Information and Knowledge Management*, pages 1983–1992.

Jindal, I., Qin, Z. T., Chen, X., Nokleby, M., and Ye, J. (2018). Optimizing taxi carpool policies via reinforcement learning and spatio-temporal mining. In *2018 IEEE International Conference on Big Data (Big Data)*, pages 1417–1426. IEEE.

Joe, W. and Lau, H. C. (2020). Deep reinforcement learning approach to solve dynamic vehicle routing problem with stochastic customers. In *Proceedings of the International Conference on Automated Planning and Scheduling*, volume 30, pages 394–402.

Kaggle (2017). Uber pickups in new york city - trip data for over 20 million uber (and other for-hire vehicle) trips in nyc. "https://www.kaggle.com/fivethirtyeight/uber-pickups-in-new-york-city".

Ke, J., Yang, H., Li, X., Wang, H., and Ye, J. (2020a). Pricing and equilibrium in on-demand ride-pooling markets. *Transportation Research Part B: Methodological*, 139:411–431.

Ke, J., Yang, H., Ye, J., et al. (2020b). Learning to delay in ride-sourcing systems: a multi-agent deep reinforcement learning framework. *IEEE Transactions on Knowledge and Data Engineering*.

Kim, S., Lewis, M. E., and White, C. C. (2005). Optimal vehicle routing with real-time traffic information. *IEEE Transactions on Intelligent Transportation Systems*, 6(2):178–188.

Kingma, D. P. and Ba, J. (2014). Adam: A method for stochastic optimization. *arXiv preprint* arXiv:1412.6980.

Kool, W., Van Hoof, H., and Welling, M. (2018). Attention, learn to solve routing problems! *arXiv preprint* arXiv:1803.08475.

Kuhn, H. W. (1955). The hungarian method for the assignment problem. *Naval research logistics quarterly*, 2(1-2):83–97.

Kullman, N. D., Cousineau, M., Goodson, J. C., and Mendoza, J. E. (2022). Dynamic ride-hailing with electric vehicles. *Transportation Science*, 56(3):775–794.

Li, L., Chu, W., Langford, J., and Schapire, R. E. (2010). A contextual-bandit approach to personalized news article recommendation. In *Proceedings of the 19th international conference on World wide web*, pages 661–670.

Li, M., Qin, Z., Jiao, Y., Yang, Y., Gong, Z., Wang, J., Wang, C., Wu, G., and Ye, J. (2019). Efficient ridesharing order dispatching with mean field multi-agent reinforcement learning. In *To appear in Proceedings of the 2019 World Wide Web Conference on World Wide Web*. International World Wide Web Conferences Steering Committee.

Li, T., Shi, C., Wang, J., Zhou, F., and Zhu, H. (2023). Optimal treatment allocation for efficient policy evaluation in sequential decision making. *arXiv preprint* arXiv:2311.02532.

Lin, B., Ghaddar, B., and Nathwani, J. (2021). Deep reinforcement learning for the electric vehicle routing problem with time windows. *IEEE Transactions on Intelligent Transportation Systems*.

Lin, K., Zhao, R., Xu, Z., and Zhou, J. (2018). Efficient large-scale fleet management via multi-agent deep reinforcement learning. In *Proceedings of the 24th ACM SIGKDD International Conference on Knowledge Discovery & Data Mining*, pages 1774–1783.

Lin, L.-J. (1992). Self-improving reactive agents based on reinforcement learning, planning and teaching. *Machine learning*, 8(3-4):293–321.

Liu, Q., Li, L., Tang, Z., and Zhou, D. (2018). Breaking the curse of horizon: Infinite-horizon off-policy estimation. In *Advances in Neural Information Processing Systems*, pages 5356–5366.

Liu, T., Xu, Z., Vignon, D., Yin, Y., Li, Q., and Qin, Z. (2023a). Effects of threshold-based incentives on drivers' labor supply behavior. *Transportation Research Part C: Emerging Technologies*, 152:104140.

Liu, T., Xu, Z., Vignon, D., Yin, Y., Qin, Z., and Li, Q. (2023b). Threshold-based incentives for ridesourcing drivers: Implications on supply management and welfare effects. *Transportation Research Part C: Emerging Technologies*, 156:104323.

Liu, Z., Li, J., and Wu, K. (2020). Context-aware taxi dispatching at city-scale using deep reinforcement learning. *IEEE Transactions on Intelligent Transportation Systems*.

Lopez, P. A., Behrisch, M., Bieker-Walz, L., Erdmann, J., Flötteröd, Y.-P., Hilbrich, R., Lücken, L., Rummel, J., Wagner, P., and Wießner, E. (2018). Microscopic traffic simulation using sumo. In *The 21st IEEE International Conference on Intelligent Transportation Systems*. IEEE.

Lowalekar, M., Varakantham, P., and Jaillet, P. (2018). Online spatio-temporal matching in stochastic and dynamic domains. *Artificial Intelligence*, 261:71–112.

Lowe, R., Wu, Y., Tamar, A., Harb, J., Abbeel, P., and Mordatch, I. (2017a). Multi-agent actor-critic for mixed cooperative-competitive environments. In *Proceedings of the 31st International Conference on Neural Information Processing Systems*, NIPS'17, page 6382-6393, Red Hook, NY, USA. Curran Associates Inc.

Lowe, R., Wu, Y. I., Tamar, A., Harb, J., Pieter Abbeel, O., and Mordatch, I. (2017b). Multi-agent actor-critic for mixed cooperative-competitive environments. *Advances in neural information processing systems*, 30.

Lyu, G., Cheung, W. C., Teo, C.-P., and Wang, H. (2019). Multi-objective online ride-matching. *Available at SSRN 3356823*.

Ma, H., Fang, F., and Parkes, D. C. (2020). Spatio-temporal pricing for ridesharing platforms. *ACM SIGecom Exchanges*, 18(2):53–57.

Mainali, M. K., Shimada, K., Mabu, S., and Hirasawa, K. (2008). Optimal route based on dynamic programming for road networks. *Journal of Advanced Computational Intelligence and Intelligent Informatics*, 12(6):546–553.

Mao, C., Liu, Y., and Shen, Z.-J. M. (2020). Dispatch of autonomous vehicles for taxi services: A deep reinforcement learning approach. *Transportation Research Part C: Emerging Technologies*, 115:102626.

Mao, C. and Shen, Z. (2018). A reinforcement learning framework for the adaptive routing problem in stochastic time-dependent network. *Transportation Research Part C: Emerging Technologies*, 93:179–197.

Mazumdar, E., Ratliff, L. J., Fiez, T., and Sastry, S. S. (2017). Gradient-based inverse risk-sensitive reinforcement learning. In *2017 IEEE 56th Annual Conference on Decision and Control (CDC)*, pages 5796–5801. IEEE.

Mehta, B., Deleu, T., Raparthy, S. C., Pal, C. J., and Paull, L. (2020). Curriculum in gradient-based meta-reinforcement learning. *arXiv preprint* arXiv:2002.07956.

Miao, F., Han, S., Lin, S., Stankovic, J. A., Zhang, D., Munir, S., Huang, H., He, T., and Pappas, G. J. (2016). Taxi dispatch with real-time sensing data in metropolitan areas: A receding horizon control approach. *IEEE Transactions on Automation Science and Engineering*, 13(2):463–478.

Miller, J. and How, J. P. (2017). Predictive positioning and quality of service ridesharing for campus mobility on demand systems. In *2017 IEEE International Conference on Robotics and Automation (ICRA)*, pages 1402–1408. IEEE.

Mnih, V., Badia, A. P., Mirza, M., Graves, A., Lillicrap, T., Harley, T., Silver, D., and Kavukcuoglu, K. (2016). Asynchronous methods for deep reinforcement learning. In *International conference on machine learning*, pages 1928–1937.

Mnih, V., Kavukcuoglu, K., Silver, D., Rusu, A. A., Veness, J., Bellemare, M. G., Graves, A., Riedmiller, M., Fidjeland, A. K., Ostrovski, G., et al. (2015). Human-level control through deep reinforcement learning. *Nature*, 518(7540):529–533.

Nazari, M., Oroojlooy, A., Snyder, L., and Takác, M. (2018). Reinforcement learning for solving the vehicle routing problem. In *Advances in Neural Information Processing Systems*, pages 9839–9849.

Ng, A. Y., Russell, S. J., et al. (2000). Algorithms for inverse reinforcement learning. In *Icml*, volume 1, page 2.

Oda, T. and Joe-Wong, C. (2018). Movi: A model-free approach to dynamic fleet management. In *IEEE INFOCOM 2018-IEEE Conference on Computer Communications*, pages 2708–2716. IEEE.

Ong, H. Y., Freund, D., and Crapis, D. (2021). Driver positioning and incentive budgeting with an escrow mechanism for ride-sharing platforms. *INFORMS Journal on Applied Analytics*, 51(5):373–390.

Oroojlooy, A. and Hajinezhad, D. (2022). A review of cooperative multi-agent deep reinforcement learning. *Applied Intelligence*, pages 1–46.

Özkan, E. and Ward, A. R. (2020). Dynamic matching for real-time ride sharing. *Stochastic Systems*, 10(1):29–70.

Pednault, E., Abe, N., and Zadrozny, B. (2002). Sequential cost-sensitive decision making with reinforcement learning. In *Proceedings of the eighth ACM SIGKDD international conference on Knowledge discovery and data mining*, pages 259–268.

Piorkowski, M., Sarafijanovic-Djukic, N., and Grossglauser, M. (2009). CRAWDAD dataset epfl/mobility (v. 2009-02-24). Downloaded from https://crawdad.org/epfl/mobility/20090224.

Pomerleau, D. A. (1991). Efficient training of artificial neural networks for autonomous navigation. *Neural computation*, 3(1):88–97.

Portal, C. D. (2020). Chicago transportation network providers (rideshare) data. "https://data.cityofchicago.org/Transportation/Transportation-Network-Providers-Trips/m6dm-c72p".

Powell, W. B. (2007). *Approximate Dynamic Programming: Solving the curses of dimensionality*, volume 703. John Wiley & Sons.

Qin, G., Luo, Q., Yin, Y., Sun, J., and Ye, J. (2021a). Optimizing matching time intervals for ride-hailing services using reinforcement learning. *Transportation Research Part C: Emerging Technologies*, 129:103239.

Qin, Z., Tang, X., Jiao, Y., Zhang, F., Xu, Z., Zhu, H., and Ye, J. (2020a). Ride-hailing order dispatching at didi via reinforcement learning. *INFORMS Journal on Applied Analytics*, 50(5):272–286.

Qin, Z., Zhu, H., and Ye, J. (2021b). Reinforcement learning for ridesharing: A survey. In *IEEE Intelligent Transportation Systems Conference*. IEEE.

Qin, Z. T., Tang, X., Zhang, L., Zhang, F., Zhang, C., Zhang, J., and Ma, Y. (2020b). *KDD Cup 2020 Reinforcement Learning Track*. https://www.kdd.org/kdd2020/kdd-cup.

Qin, Z. T., Zhu, H., and Ye, J. (2022). Reinforcement learning for ridesharing: An extended survey. *Transportation Research Part C: Emerging Technologies*, 144:103852.

Raju, C., Narahari, Y., and Ravikumar, K. (2003). Reinforcement learning applications in dynamic pricing of retail markets. In *IEEE International Conference on E-Commerce, 2003. CEC 2003.*, pages 339–346. IEEE.

Ramos, G. d. O., Bazzan, A. L., and da Silva, B. C. (2018). Analysing the impact of travel information for minimising the regret of route choice. *Transportation Research Part C: Emerging Technologies*, 88:257–271.

Riley, C., Van Hentenryck, P., and Yuan, E. (2020). Real-time dispatching of large-scale ride-sharing systems: Integrating optimization, machine learning, and model predictive control. *Proceedings of the Twenty-Ninth International Joint Conference on Artificial Intelligence (IJCAI)*.

Rong, H., Zhou, X., Yang, C., Shafiq, Z., and Liu, A. (2016). The rich and the poor: A markov decision process approach to optimizing taxi driver revenue efficiency. In *Proceedings of the 25th ACM International on Conference on Information and Knowledge Management*, pages 2329–2334.

Ross, S., Gordon, G., and Bagnell, D. (2011). A reduction of imitation learning and structured prediction to no-regret online learning. In *Proceedings of the fourteenth international conference on artificial intelligence and statistics*, pages 627–635. JMLR Workshop and Conference Proceedings.

Ruch, C., Hörl, S., and Frazzoli, E. (2018). Amodeus, a simulation-based testbed for autonomous mobility-on-demand systems. In *2018 21st International Conference on Intelligent Transportation Systems (ITSC)*, pages 3639–3644. IEEE.

Russell, S. (1998). Learning agents for uncertain environments. In *Proceedings of the eleventh annual conference on Computational learning theory*, pages 101–103.

Rusu, A. A., Rabinowitz, N. C., Desjardins, G., Soyer, H., Kirkpatrick, J., Kavukcuoglu, K., Pascanu, R., and Hadsell, R. (2016). Progressive neural networks. *arXiv preprint* arXiv:1606.04671.

Sahr, K. (2011). Hexagonal discrete global grid systems for geospatial computing. *Archiwum Fotogrametrii, Kartografii i Teledetekcji*, 22:363–376.

Schmoll, S. and Schubert, M. (2020). Semi-markov reinforcement learning for stochastic resource collection. *Proceedings of the Twenty-Ninth International Joint Conference on Artificial Intelligence (IJCAI)*.

Schulman, J., Levine, S., Abbeel, P., Jordan, M., and Moritz, P. (2015). Trust region policy optimization. In *International conference on machine learning*, pages 1889–1897. PMLR.

Schulman, J., Wolski, F., Dhariwal, P., Radford, A., and Klimov, O. (2017). Proximal policy optimization algorithms. *arXiv preprint* arXiv:1707.06347.

Shah, S., Lowalekar, M., and Varakantham, P. (2020). Neural approximate dynamic programming for on-demand ride-pooling. In *Proceedings of the AAAI Conference on Artificial Intelligence*, volume 34, pages 507–515.

Shang, W., Li, Q., Qin, Z., Yu, Y., Meng, Y., and Ye, J. (2021). Partially observable environment estimation with uplift inference for reinforcement learning based recommendation. *Machine Learning*, pages 1–38.

Shang, W., Yu, Y., Li, Q., Qin, Z., Meng, Y., and Ye, J. (2019). Environment reconstruction with hidden confounders for reinforcement learning based recommendation. In *Proceedings of the 25th ACM SIGKDD International Conference on Knowledge Discovery & Data Mining*, pages 566–576.

Shen, W., He, X., Zhang, C., Ni, Q., Dou, W., and Wang, Y. (2020). Auxiliary-task based deep reinforcement learning for participant selection problem in mobile crowdsourcing. In *Proceedings of the 29th ACM International Conference on Information & Knowledge Management*, pages 1355–1364.

Shi, C., Wan, R., Song, G., Luo, S., Zhu, H., and Song, R. (2023a). A multiagent reinforcement learning framework for off-policy evaluation in two-sided markets. *The Annals of Applied Statistics*, 17(4):2701–2722.

Shi, C., Wang, X., Luo, S., Zhu, H., Ye, J., and Song, R. (2023b). Dynamic causal effects evaluation in a/b testing with a reinforcement learning framework. *Journal of the American Statistical Association*, 118(543):2059–2071.

Shi, C., Zhu, J., Ye, S., Luo, S., Zhu, H., and Song, R. (2022). Off-policy confidence interval estimation with confounded markov decision process. *Journal of the American Statistical Association*, pages 1–12.

Shi, J., Gao, Y., Wang, W., Yu, N., and Ioannou, P. A. (2019a). Operating electric vehicle fleet for ride-hailing services with reinforcement learning. *IEEE Transactions on Intelligent Transportation Systems*, 21(11):4822–4834.

Shi, J.-C., Yu, Y., Da, Q., Chen, S.-Y., and Zeng, A.-X. (2019b). Virtual-taobao: Virtualizing real-world online retail environment for reinforcement learning. In *Proceedings of the AAAI Conference on Artificial Intelligence*, volume 33, pages 4902–4909.

Shou, Z. and Di, X. (2020a). Multi-agent reinforcement learning for dynamic routing games: A unified paradigm. *arXiv preprint* arXiv:2011.10915.

Shou, Z. and Di, X. (2020b). Reward design for driver repositioning using multi-agent reinforcement learning. *Transportation research part C: emerging technologies*, 119:102738.

Shou, Z., Di, X., Ye, J., Zhu, H., Zhang, H., and Hampshire, R. (2020). Optimal passenger-seeking policies on e-hailing platforms using markov decision process and imitation learning. *Transportation Research Part C: Emerging Technologies*, 111:91–113.

Silver, D. and Hassabis, D. (2016). Alphago: Mastering the ancient game of go with machine learning. *Research Blog*, 9.

Simao, H. P., Day, J., George, A. P., Gifford, T., Nienow, J., and Powell, W. B. (2009). An approximate dynamic programming algorithm for large-scale fleet management: A case application. *Transportation Science*, 43(2):178–197.

Singh, A., Al-Abbasi, A. O., and Aggarwal, V. (2021). A distributed model-free algorithm for multi-hop ride-sharing using deep reinforcement learning. *IEEE Transactions on Intelligent Transportation Systems*.

Smith, M. (2019). *Here?s how long you have to wait for an Uber or Lyft in DC*. https://wtop.com/dc-transit/2019/12/how-long-you-have-to-wait-for-an-uber-or-lyft-in-d-c/.

Song, J., Cho, Y. J., Kang, M. H., and Hwang, K. Y. (2020). An application of reinforced learning-based dynamic pricing for improvement of ridesharing platform service in seoul. *Electronics*, 9(11):1818.

Sun, H., Wang, H., and Wan, Z. (2019). Model and analysis of labor supply for ride-sharing platforms in the presence of sample self-selection and endogeneity. *Transportation Research Part B: Methodological*, 125:76–93.

Sutton, R. S. (1988). Learning to predict by the methods of temporal differences. *Machine learning*, 3(1):9–44.

Sutton, R. S. and Barto, A. G. (2018). *Reinforcement learning: An introduction*. MIT press.

Sutton, R. S., Barto, A. G., et al. (1998). *Reinforcement learning: An introduction*. MIT press.

Sutton, R. S., Precup, D., and Singh, S. (1999). Between mdps and semi-mdps: A framework for temporal abstraction in reinforcement learning. *Artificial intelligence*, 112(1-2):181–211.

Tampuu, A., Matiisen, T., Kodelja, D., Kuzovkin, I., Korjus, K., Aru, J., Aru, J., and Vicente, R. (2017). Multiagent cooperation and competition with deep reinforcement learning. *PLOS ONE*, 12(4):1–15.

Tang, X., Qin, Z., Zhang, F., Wang, Z., Xu, Z., Ma, Y., Zhu, H., and Ye, J. (2019). A deep value-network based approach for multi-driver order dispatching. In *Proceedings of the 25th ACM SIGKDD international conference on knowledge discovery & data mining*, pages 1780–1790.

Tang, X., Zhang, F., Qin, Z., Wang, Y., Shi, D., Song, B., Tong, Y., Zhu, H., and Ye, J. (2021). Value function is all you need: A unified learning framework for ride hailing platforms. In *Proceedings of the 27th ACM SIGKDD Conference on Knowledge Discovery & Data Mining*, KDD '21, page 3605-3615, New York, NY, USA. Association for Computing Machinery.

Theocharous, G., Thomas, P. S., and Ghavamzadeh, M. (2015). Ad recommendation systems for life-time value optimization. In *Proceedings of the 24th international conference on world wide web*, pages 1305–1310.

TLC (2020). Nyc taxi & limousine commission trip record data. "https://www1.nyc.gov/site/tlc/about/tlc-trip-record-data.page".

Tong, Y., Shi, D., Xu, Y., Lv, W., Qin, Z., and Tang, X. (2021). Combinatorial optimization meets reinforcement learning: Effective taxi order dispatching at large-scale. *IEEE Transactions on Knowledge and Data Engineering*.

Tong, Y., Zeng, Y., Zhou, Z., Chen, L., Ye, J., and Xu, K. (2018). A unified approach to route planning for shared mobility. *Proceedings of the VLDB Endowment*, 11(11):1633.

Tong, Y., Zhou, Z., Zeng, Y., Chen, L., and Shahabi, C. (2020). Spatial crowdsourcing: a survey. *The VLDB Journal*, 29(1):217–250.

Traoré, R., Caselles-Dupré, H., Lesort, T., Sun, T., Díaz-Rodríguez, N., and Filliat, D. (2019). Continual reinforcement learning deployed in real-life using policy distillation and sim2real transfer. *arXiv preprint* arXiv:1906.04452.

Truong, C., Oudre, L., and Vayatis, N. (2020). Selective review of offline change point detection methods. *Signal Processing*, 167:107299.

Tumer, K., Welch, Z. T., and Agogino, A. (2008). Aligning social welfare and agent preferences to alleviate traffic congestion. In *Proceedings of the 7th international joint conference on Autonomous agents and multiagent systems-Volume 2*, pages 655–662. Citeseer.

Turan, B., Pedarsani, R., and Alizadeh, M. (2020). Dynamic pricing and fleet management for electric autonomous mobility on demand systems. *Transportation Research Part C: Emerging Technologies*, 121:102829.

Uber (2021). Uber movement. "https://movement.uber.com/?lang=en-US".

Uher, V., Gajdoš, P., Snášel, V., Lai, Y.-C., and Radecký, M. (2019). Hierarchical hexagonal clustering and indexing. *Symmetry*, 11(6):731.

Ulmer, M. W., Goodson, J. C., Mattfeld, D. C., and Thomas, B. W. (2020). On modeling stochastic dynamic vehicle routing problems. *EURO Journal on Transportation and Logistics*, 9(2):100008.

Urata, J., Xu, Z., Ke, J., Yin, Y., Wu, G., Yang, H., and Ye, J. (2021). Learning ride-sourcing drivers' customer-searching behavior: A dynamic discrete choice approach. *Transportation Research Part C: Emerging Technologies*, 130:103293.

Van Hasselt, H., Guez, A., and Silver, D. (2016). Deep reinforcement learning with double q-learning. In *AAAI*, pages 2094–2100.

Verma, T., Varakantham, P., Kraus, S., and Lau, H. C. (2017). Augmenting decisions of taxi drivers through reinforcement learning for improving revenues. In *Twenty-Seventh International Conference on Automated Planning and Scheduling*.

Vinyals, O., Fortunato, M., and Jaitly, N. (2015). Pointer networks. In *Advances in Neural Information Processing Systems*, pages 2692–2700.

W Axhausen, K., Horni, A., and Nagel, K. (2016). *The multi-agent transport simulation MATSim*. Ubiquity Press.

Wang, H. and Yang, H. (2019). Ridesourcing systems: A framework and review. *Transportation Research Part B: Methodological*, 129:122–155.

Wang, Y., Tong, Y., Long, C., Xu, P., Xu, K., and Lv, W. (2019). Adaptive dynamic bipartite graph matching: A reinforcement learning approach. In *2019 IEEE 35th International Conference on Data Engineering (ICDE)*, pages 1478–1489. IEEE.

Wang, Z., Bapst, V., Heess, N., Mnih, V., Munos, R., Kavukcuoglu, K., and de Freitas, N. (2016). Sample efficient actor-critic with experience replay. *arXiv preprint* arXiv:1611.01224.

Wang, Z., Qin, Z., Tang, X., Ye, J., and Zhu, H. (2018). Deep reinforcement learning with knowledge transfer for online rides order dispatching. In *International Conference on Data Mining*. IEEE.

Watkins, C. J. and Dayan, P. (1992). Q-learning. *Machine learning*, 8(3-4):279–292.

Wei, H., Yang, Z., Liu, X., Qin, Z., Tang, X., and Ying, L. (2023). A reinforcement learning and prediction-based lookahead policy for vehicle repositioning in online ride-hailing systems. *IEEE Transactions on Intelligent Transportation Systems*.

Wen, F., Wang, X., and Xu, X. (2019). Hierarchical sarsa learning based route guidance algorithm. *Journal of Advanced Transportation*, 2019.

Wen, J., Zhao, J., and Jaillet, P. (2017). Rebalancing shared mobility-on-demand systems: A reinforcement learning approach. In *2017 IEEE 20th International Conference on Intelligent Transportation Systems (ITSC)*, pages 220–225. Ieee.

Williams, R. J. (1992). Simple statistical gradient-following algorithms for connectionist reinforcement learning. *Machine learning*, 8(3-4):229–256.

Wong, R., Szeto, W., and Wong, S. (2014). A cell-based logit-opportunity taxi customer-search model. *Transportation Research Part C: Emerging Technologies*, 48:84–96.

Wu, C., Kreidieh, A., Parvate, K., Vinitsky, E., and Bayen, A. M. (2017). Flow: Architecture and benchmarking for reinforcement learning in traffic control. *arXiv preprint* arXiv:1710.05465, page 10.

Wu, T., Joseph, A. D., and Russell, S. J. (2016). Automated pricing agents in the on-demand economy. *University of California at Berkeley: Berkeley, CA, USA*.

Wu, Y., Li, Q., and Qin, Z. (2022). Spatio-temporal incentives optimization for ride-hailing services with offline deep reinforcement learning. *arXiv preprint* arXiv:2211.03240.

Xu, Y., Tong, Y., Shi, Y., Tao, Q., Xu, K., and Li, W. (2020). An efficient insertion operator in dynamic ridesharing services. *IEEE Transactions on Knowledge and Data Engineering*.

Xu, Z., Li, Z., Guan, Q., Zhang, D., Li, Q., Nan, J., Liu, C., Bian, W., and Ye, J. (2018). Large-scale order dispatch in on-demand ride-hailing platforms: A learning and planning approach. In *Proceedings of the 24th ACM SIGKDD International Conference on Knowledge Discovery & Data Mining*, pages 905–913. ACM.

Yan, C., Zhu, H., Korolko, N., and Woodard, D. (2020). Dynamic pricing and matching in ride-hailing platforms. *Naval Research Logistics (NRL)*, 67(8):705–724.

Yang, H., Qin, X., Ke, J., and Ye, J. (2020a). Optimizing matching time interval and matching radius in on-demand ride-sourcing markets. *Transportation Research Part B: Methodological*, 131:84–105.

Yang, H., Shao, C., Wang, H., and Ye, J. (2020b). Integrated reward scheme and surge pricing in a ridesourcing market. *Transportation Research Part B: Methodological*, 134:126–142.

Yang, Y., Luo, R., Li, M., Zhou, M., Zhang, W., and Wang, J. (2018). Mean field multi-agent reinforcement learning. In *International Conference on Machine Learning*, pages 5571–5580. PMLR.

Yao, R. and Bekhor, S. (2021). A ridesharing simulation platform that considers dynamic supply-demand interactions. *arXiv preprint* arXiv:2104.13463.

Yau, K.-L. A., Qadir, J., Khoo, H. L., Ling, M. H., and Komisarczuk, P. (2017). A survey on reinforcement learning models and algorithms for traffic signal control. *ACM Comput. Surv.*, 50(3).

Yu, S., Zhou, J., Li, B., Mabu, S., and Hirasawa, K. (2012). Q value-based dynamic programming with sarsa learning for real time route guidance in large scale road networks. In *The 2012 International Joint Conference on Neural Networks (IJCNN)*, pages 1–7. IEEE.

Yu, X., Gao, S., Hu, X., and Park, H. (2019). A markov decision process approach to vacant taxi routing with e-hailing. *Transportation Research Part B: Methodological*, 121:114–134.

Yu, X. and Shen, S. (2019). An integrated decomposition and approximate dynamic programming approach for on-demand ride pooling. *IEEE Transactions on Intelligent Transportation Systems*, 21(9):3811–3820.

Yuen, C. F., Singh, A. P., Goyal, S., Ranu, S., and Bagchi, A. (2019). Beyond shortest paths: Route recommendations for ride-sharing. In *The World Wide Web Conference*, pages 2258–2269.

Zeng, Y., Cai, R., Sun, F., Huang, L., and Hao, Z. (2023). A survey on causal reinforcement learning. *arXiv preprint* arXiv:2302.05209.

Zhang, K., He, F., Zhang, Z., Lin, X., and Li, M. (2020a). Multi-vehicle routing problems with soft time windows: A multi-agent reinforcement learning approach. *Transportation Research Part C: Emerging Technologies*, 121:102861.

Zhang, L., Hu, T., Min, Y., Wu, G., Zhang, J., Feng, P., Gong, P., and Ye, J. (2017). A taxi order dispatch model based on combinatorial optimization. In *Proceedings of the 23rd ACM SIGKDD International Conference on Knowledge Discovery and Data Mining*, pages 2151–2159. ACM.

Zhang, R., Rossi, F., and Pavone, M. (2016). Model predictive control of autonomous mobility-on-demand systems. In *2016 IEEE International Conference on Robotics and Automation (ICRA)*, pages 1382–1389. IEEE.

Zhang, W., Wang, Q., Li, J., and Xu, C. (2020b). Dynamic fleet management with rewriting deep reinforcement learning. *IEEE Access*, 8:143333–143341.

Zhao, X., Zhang, L., Ding, Z., Xia, L., Tang, J., and Yin, D. (2018). Recommendations with negative feedback via pairwise deep reinforcement learning. In *Proceedings of the 24th ACM SIGKDD International Conference on Knowledge Discovery & Data Mining*, pages 1040–1048.

Zheng, L., Chen, L., and Ye, J. (2018). Order dispatch in price-aware ridesharing. *Proceedings of the VLDB Endowment*, 11(8):853–865.

Zhong, Y., Wan, Z., and Shen, Z.-J. M. (2020). Queueing versus surge pricing mechanism: Efficiency, equity, and consumer welfare. *Equity, and Consumer Welfare (September 24, 2020)*.

Zhou, B., Song, Q., Zhao, Z., and Liu, T. (2020a). A reinforcement learning scheme for the equilibrium of the in-vehicle route choice problem based on congestion game. *Applied Mathematics and Computation*, 371:124895.

Zhou, F., Lu, C., Tang, X., Zhang, F., Qin, Z., Ye, J., and Zhu, H. (2021a). Multi-objective distributional reinforcement learning for large-scale order dispatching. In *2021 IEEE International Conference on Data Mining (ICDM)*, pages 1541–1546. IEEE.

Zhou, F., Luo, S., Qie, X., Ye, J., and Zhu, H. (2021b). Graph-based equilibrium metrics for dynamic supply–demand systems with applications to ride-sourcing platforms. *Journal of the American Statistical Association*, pages 1–12.

Zhou, M., Jin, J., Zhang, W., Qin, Z., Jiao, Y., Wang, C., Wu, G., Yu, Y., and Ye, J. (2019). Multi-agent reinforcement learning for order-dispatching via order-vehicle distribution matching. In *Proceedings of the 28th ACM International Conference on Information and Knowledge Management*, pages 2645–2653.

Zhou, M., Luo, J., Villela, J., Yang, Y., Rusu, D., Miao, J., Zhang, W., Alban, M., Fadakar, I., Chen, Z., et al. (2020b). Smarts: Scalable multi-agent reinforcement learning training school for autonomous driving. *arXiv preprint* arXiv:2010.09776.

Zhou, X., Rong, H., Yang, C., Zhang, Q., Khezerlou, A. V., Zheng, H., Shafiq, Z., and Liu, A. X. (2018). Optimizing taxi driver profit efficiency: A spatial network-based markov decision process approach. *IEEE Transactions on Big Data*, 6(1):145–158.

Zhou, Y., Yang, H., Ke, J., Wang, H., and Li, X. (2022). Competition and third-party platform-integration in ride-sourcing markets. *Transportation Research Part B: Methodological*, 159:76–103.

Zhu, Z., Ke, J., and Wang, H. (2021). A mean-field markov decision process model for spatial-temporal subsidies in ride-sourcing markets. *Transportation Research Part B: Methodological*, 150:540–565.

Zou, F., Yen, G. G., and Zhao, C. (2021). Dynamic multiobjective optimization driven by inverse reinforcement learning. *Information Sciences*, 575:468–484.